Neuroscience: a Primer

Written and Designed by Students in the Neuroscience Graduate Group at the Perelman School of Medicine at the University of Pennsylvania

Edited by Lauren Stutzbach

Copyright © 2014 by Lauren Stutzbach

All rights reserved. This book or any portion thereof may not be reproduced or used in any manner whatsoever without the express written permission of the publisher except for the use of brief quotations in a book review or scholarly journal.

This work is licensed under the Creative Commons Attribution-ShareAlike 4.0 International License. To view a copy of this license, visit http://creativecommons.org/licenses/by-sa/4.0/ or send a letter to Creative Commons, PO Box 1866, Mountain View, CA 94042, USA.

First Printing: 2014

ISBN 978-1-329-06245-0

Neuroscience Graduate Group, University of Pennsylvania
140 John Morgan Building
3620 Hamilton Walk
Philadelphia, PA 19104

www.med.upenn.edu/ngg
www.knowyourmind.org

Contributors

Chief Editor: Lauren Stutzbach

Managing Editors:
Danielle Mor
Anna Stern
Vanessa Troiani

Copy and Figure Editors:
Elaine Liu
Trishala Parthasarathi
Shivon Robinson
M. Morgan Taylor
Khaing Win

Authors:
Jesse Isaacman-Beck
Britter Gunderson
Maria Lim
Vanessa Troiani
Jason Wester
Michelle Dumoulin
Nina Hsu
Emilia Moscato
Leif Vigeland
Samantha White
Keith Feigenson
Katie Kopil
Lauren Stutzbach
Cristin Welle
Mathieu Wimmer

Figure Designers:
Christopher Dengler
Kaitlin Folweiler
Angela Jablonski
Yin Li
Andrew Moore
Shivon Robinson
M. Morgan Taylor
Michelle Dumoulin
Adam Gifford
Blake Kimmey
Elaine Liu
Patricia Murphy
Srihari Sritharan
Lindsay Vass
Khaing Win
Yunshu Fan
Leonardo Guercio
Kim Krisada
Andrew Moberly
David Reiner
Lauren Stutzbach
Mathieu Wimmer

Cover Art: Greg A. Dunn, copyright 2009-2011, all rights reserved.
www.gregadunn.com

Cover Design: Kate Christison-Lagay

Chapter Art: Collin Challis

With generous support from the Mahoney Institute for Neurosciences,
Perelman School of Medicine at the University of Pennsylvania

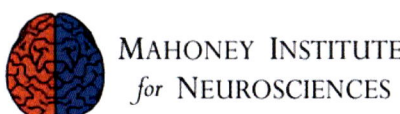

Table of Contents

Chapter 1: Introduction to Neuroscience 7

Chapter 2: The Neuron 15

Chapter 3: Electrical and Chemical Communication 23

Chapter 4: Sensation and Perception 35

Chapter 5: Learning and Memory 47

Chapter 6: Sleep 55

Chapter 7: Stress 67

Chapter 8: Aging 77

Chapter 9: Addiction 87

INTRODUCTION
UPENN NEUROSCIENCE

Chapter 1: Introduction to Neuroscience

I. What is Neuroscience?

It is pretty amazing to be a human being. Think about it: you have thoughts, emotions, dreams, and memories. For thousands of years, people have wondered what it means to be human and how it is that we are able to think and feel. Scientists, philosophers, and religious leaders have argued about these questions, but despite their efforts, we are still far from being able to answer fundamental questions about what makes humans special.

The goal of this book is to introduce you to an exciting field called **neuroscience**. So what is neuroscience exactly? It is the study of the brain, the source of all the things that make you human. Throughout this book you will learn what we know about the brain, but keep in mind that there is still plenty left to discover!

II. The Nervous System

The brain is part of a complicated network called the **nervous system**. The nervous system has two major divisions: 1) the **central nervous system**, which is made up of the **brain** in your head and the **spinal cord** in your back, and 2) the **peripheral nervous system**, which is made up of all the **nerves** connecting your spinal cord to all the organs and muscles in your body. Figure 1-1 shows a diagram of these two divisions of the nervous system.

Think of the brain as the boss of the nervous system. It's the

---

Neuroscience: the study of the brain

Nervous System: a body system that consists of brain, spinal cord, and nerves. This system enables the organism to sense the environment and respond appropriately. It controls movement, sensation, thoughts, and emotions, and it regulates all of the body's functions.

Central nervous system (CNS): the part of the nervous system that is composed of the brain and spinal cord

Brain: an organ of the nervous system that acts as the control center, governing mental processes such as thinking, feeling, processing sensation, and directing movement

Spinal Cord: a component of the nervous system that is located inside the spinal column, or backbone. It helps to send messages between the brain and the rest of the body.

control center that processes all the information from the environment that you pick up through your ears, eyes, nose, mouth, and skin. The brain uses this information to tell your body to respond appropriately. The spinal cord is like the brain's assistant, communicating messages between the brain and the body. The spinal cord uses nerves like telephone wires to deliver messages to and receive messages from the skin, muscles, and other organs. For example, when your brain decides that you want a snack, it tells the spinal cord. The spinal cord then activates the nerves in your arm to make you pick up a banana. This may sound simple, but it's actually a pretty complex process!

Peripheral nervous system (PNS): the part of the nervous system that is composed of nerves that communicate with muscles and organs

Nerves: the connections between the spinal cord and the rest of the body

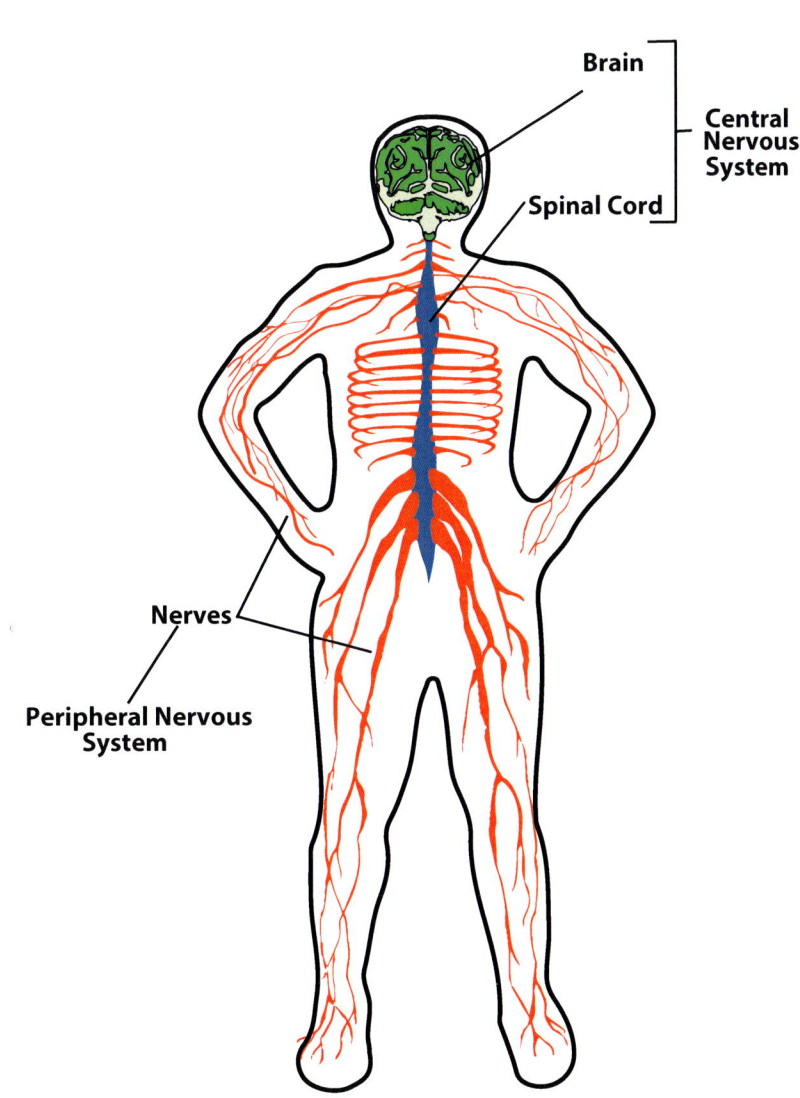

Figure 1-1: *Divisions of the nervous system*

Cortex: the outermost layer of the brain responsible for most complex functions. It is divided into four lobes.

Frontal Lobe: the part of the cortex that is responsible for judgment, movement, and emotions

Parietal Lobe: the part of the cortex that is responsible for processing sensory information, especially touch

Occipital Lobe: the part of the cortex that is responsible for vision

Together, the parts of the nervous system govern everything your body does. Some of these actions require you to think and focus (like reading, texting, and flirting), while others happen without your awareness (like breathing, sleeping, and digesting). Whether you're paying attention or not, your nervous system keeps you going.

III. The Structures of the Brain

When you first look at the brain, it might seem like just a big gray blob. However, even this blob (much like the Earth) has ordered geography. Like our planet, the brain is split into two halves called hemispheres. On both hemispheres, the outer layer is called the **cortex**, and this is where the most complex functions are carried out, like consciousness. The cortex is divided into four sections, or lobes:

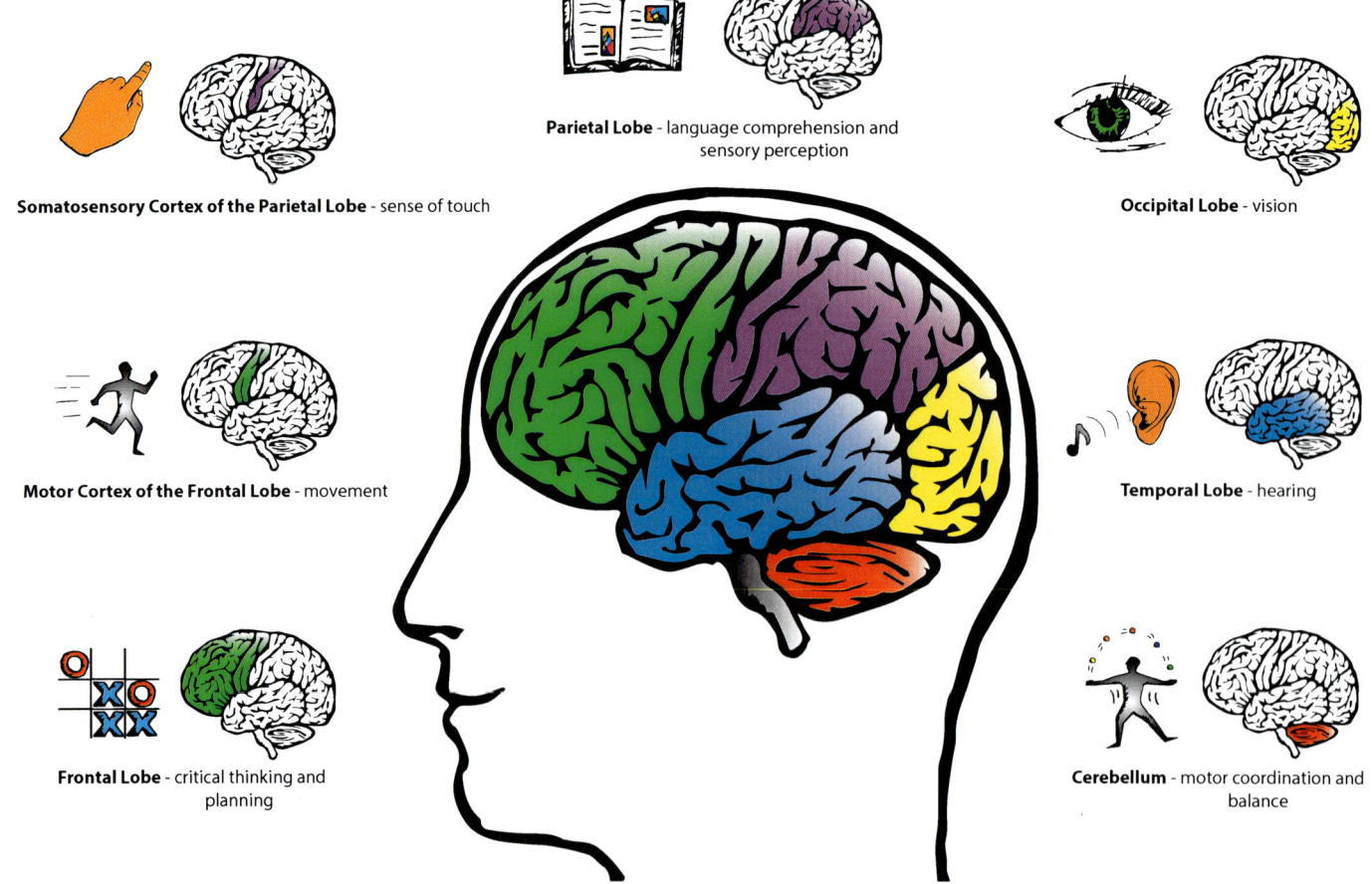

Figure 1-2: Lobes of the brain and their functions

10

frontal, **parietal**, **occipital**, and **temporal**. To see where each lobe is located on the brain, see Figure 1-2.

The frontal lobe controls judgment, movement, and emotions. The parietal lobe processes information from the different sensory systems, especially touch. The temporal lobe is important for hearing, memory, and speech. The occipital lobe is dedicated to vision. The surface of the cortex has lots of ridges because it is folded in on itself like a crumpled piece of paper. The ridges are made up of peaks (called **gyri**) and valleys (called **sulci**).

Underneath the cortex, deeper in the brain, are many other structures that take care of lots of other important functions. You will learn more about these structures throughout this book. At the bottom and back of the brain is the **cerebellum**, which calculates exactly how to carry out a movement so that the muscles can act with precise timing and coordination. The cerebellum is highlighted in red in Figure 1-2. The brain is connected to the spinal cord by the **brainstem**, which is responsible for bodily functions we are normally not aware of, such as breathing and heart rate. The brainstem is the gray structure at the bottom of the brain in Figure 1-2.

Interestingly, throughout evolution, the brain and nervous system changed a lot. As different species evolved, the brain increased in both size and complexity. For a comparison of the brains of different species, see Figure 1-3. One of the most striking differences between species is in the amount of cortex, the outer layer of the brain that is so important for consciousness. The smarter the animal, the bigger the cortex is in relation to the rest of the brain. This and other differences in brain anatomy allowed humans to become unique in their remarkable abilities to use language, to feel, and to think.

IV. The Cells of the Nervous System

How does the brain handle all of these tasks and manage to keep everything straight? Like all of the other organs in your body, the brain is made up of individual cells...lots of them! The most important type of cell in the brain is called the neuron.

> **Temporal Lobe**: the part of the cortex that is responsible for hearing, memory, and speech

> **Gyri**: "peaks" in brain tissue that make up the cortex

> **Sulci**: "valleys" in brain tissue that make up the cortex

> **Cerebellum**: a part of the brain that is important for motor control and coordination

> **Brainstem**: a part of the brain that connects with the spinal cord. It plays an important role in regulating breathing and heart rate.

> **Neuron**: a type of cell in the nervous system that uses electricity and chemicals to communicate. There are over 100 billion neurons in the brain and spinal cord.

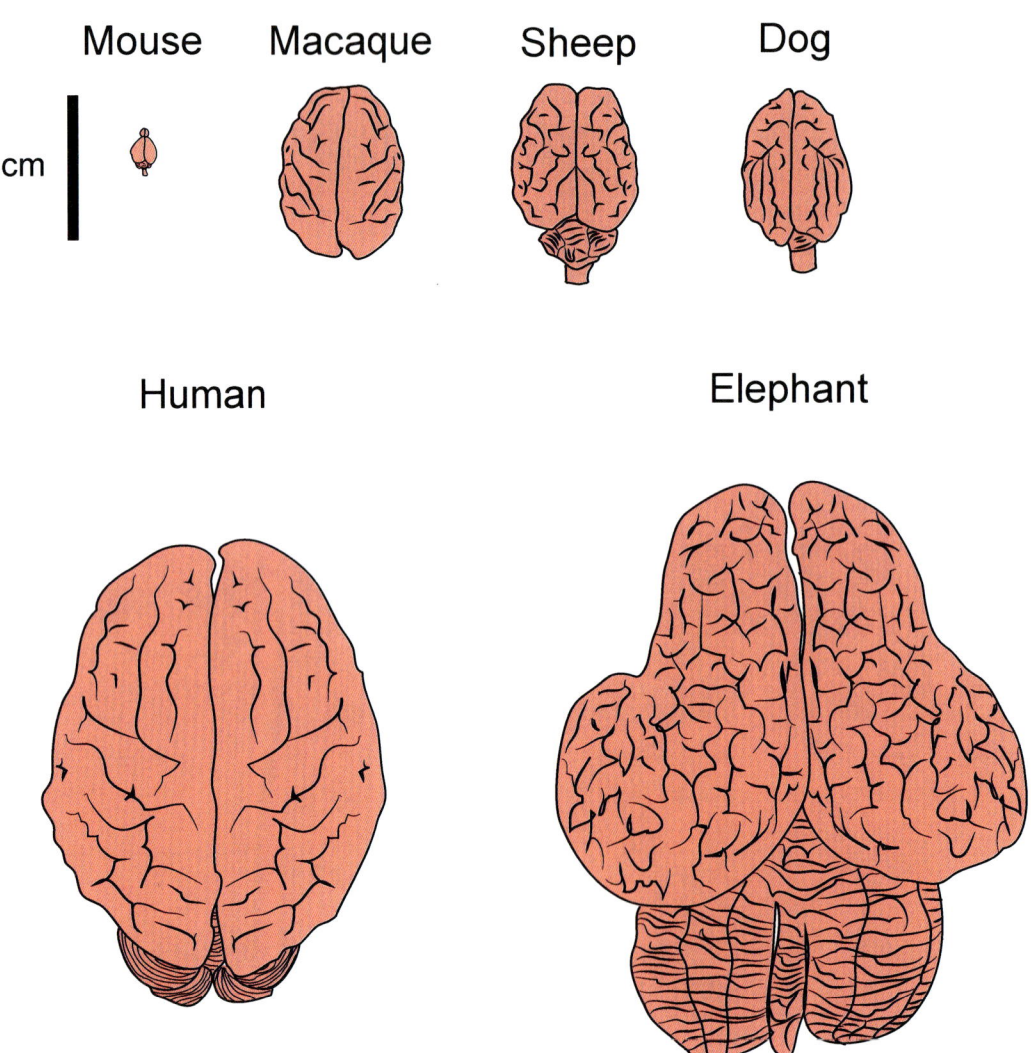

Figure 1-3: Comparison of different mammal brains

Neurons communicate using both electricity and chemicals, and they can send information to each other at lightning speed. You will learn more about how neurons communicate in Chapter 3. This speed of communication is pretty fortunate because it's the reason that when you see a lion running towards you, you're off and running in the other direction before you can even think, "AHHH!" There are over 100,000,000,000 (100 billion) neurons in your nervous system, and they form trillions (yes, trillions!) of very specific connections with one another in an effort to keep you alive and healthy (instead of, say, eaten by that lion).

Neurons are not the only cells in the nervous system. They need help doing all their hard work, and so they are surrounded by a network of supporting cells called **glia**. Glia work to keep the neurons in the central and peripheral nervous systems healthy

> **Glia**: a type of cell in the nervous system that provides support for neurons and helps them survive

and happy. There are several different types of glial cells and they each perform an important function, like delivering important nutrients to neurons, cleaning up waste from neurons, and making sure neurons are communicating at the right speed. You will learn much more about neurons and glia in the next chapter.

Throughout this textbook, you will learn about how these cells allow the brain to sense the environment and make decisions. We'll cover topics like learning and memory, what happens to your brain when you're sleeping, and even why people can develop addictions. Have fun learning about your amazing brain!

Study Questions

1. Describe how the brain, spinal cord, and nerves work together.

2. Name the four lobes of the cortex and their functions.

Challenge Question

Why do you think the cortex has so many folds?

Brain Byte

Have you ever wondered how your brain developed when you were just a baby? Believe it or not, 250,000 neurons formed per minute to create trillions of connections before you were even born! At birth, every neuron in the cortex has around 2,500 connections. By the time you're two or three years old, every neuron has around 15,000 connections to other neurons! This is more than twice the number of connections in the average adult brain. As you grow older, your brain keeps the connections it uses a lot and destroys the connections it uses only a little or not at all. People once believed that the number of connections doesn't change after a certain time in your life, but we've recently discovered that the brain is constantly changing and adjusting itself.

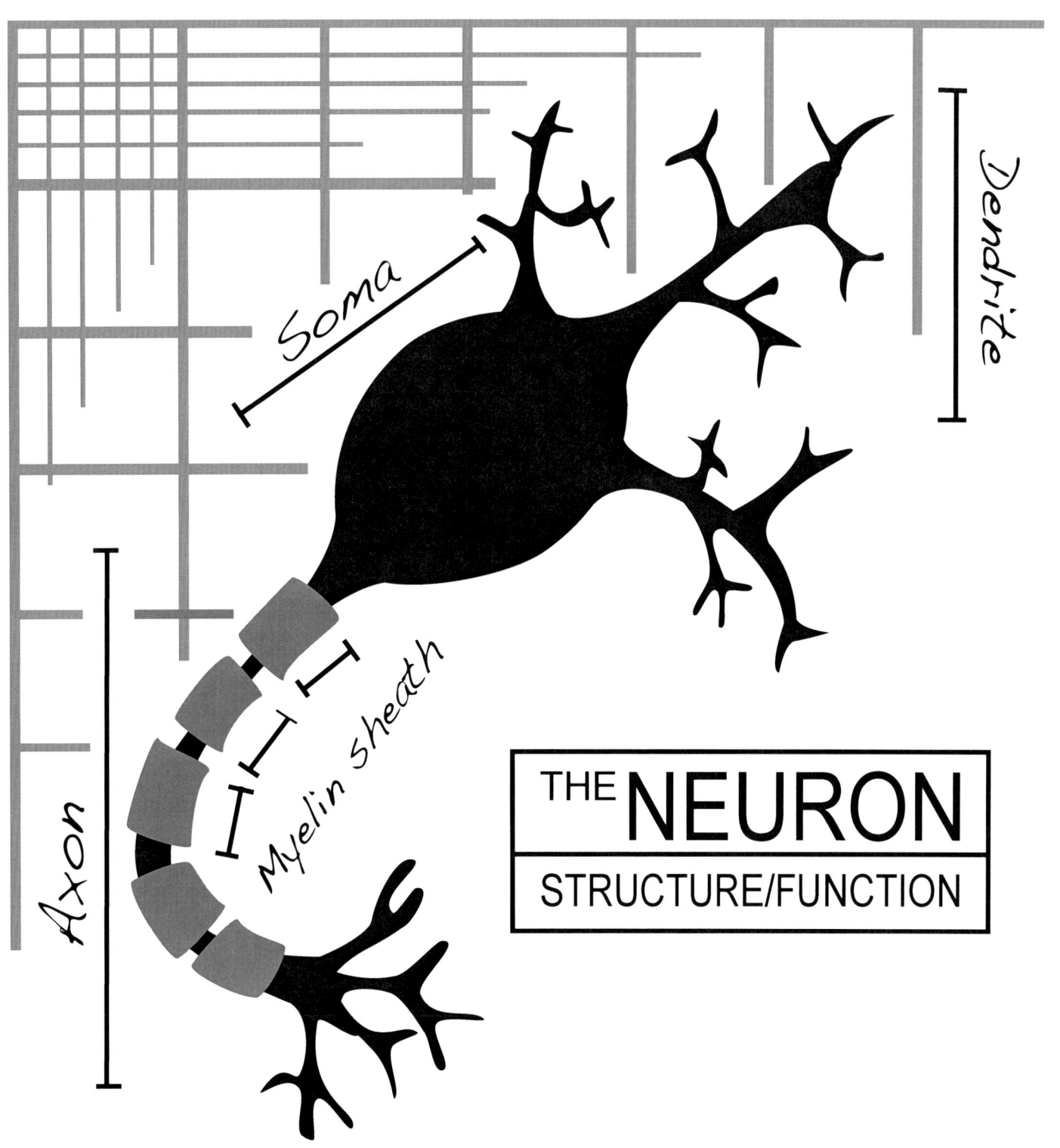

Chapter 2: The Neuron

I. What are the cells in the Nervous System?

IA. Structure of the Neuron

As you learned in Chapter 1, the main cells that make up the nervous system are called neurons. Neurons have a unique structure that makes them different from all the other cells in the body. This is because neurons have a very special job: communication! Let's find out how a neuron's shape, also called its structure, gives it the ability to communicate with other neurons and with the rest of the body.

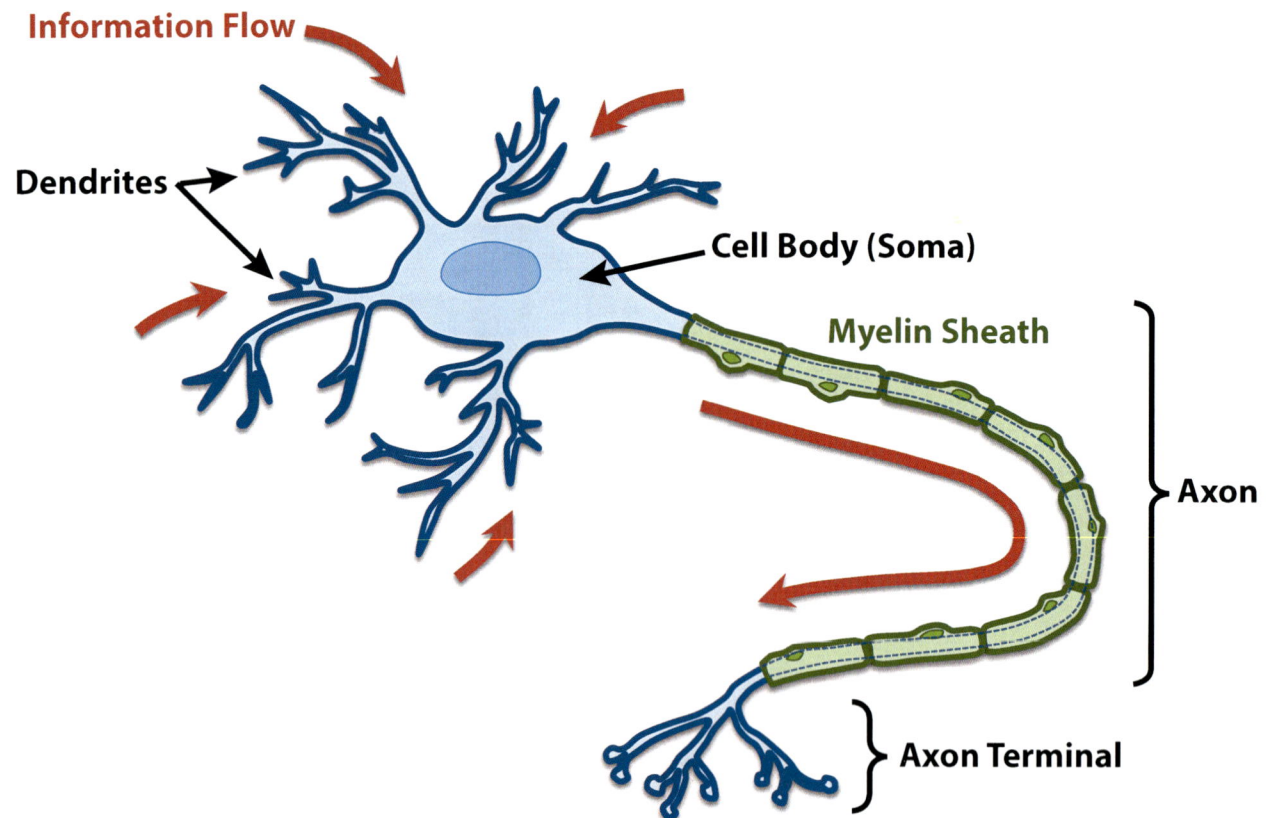

Figure 2-1: *The parts of a neuron (labels in black and green) and the flow of information through the neuron (red arrows)*

The neuron has three main parts: **dendrites**, a **cell body**, and an **axon**. For a diagram of this, see Figure 2-1. The dendrites are branch-like structures at one end of the neuron. They receive incoming signals from other neurons. The cell body, also called the soma, is home to the neuron's nucleus and all of its organelles, which it needs in order to stay alive. The cell body processes the signals coming in through the dendrites and uses this information to determine what message will be sent to the next neuron. This new message is sent to the axon, which is a long, thin structure that carries the signal extremely quickly all the way to the **axon terminal**. The axon terminal acts as an outbox for the neuron, sending a signal to the dendrites of the next neuron. You will find out more about how exactly this happens in Chapter 3.

Axons can be very long, allowing neurons to communicate with other neurons very far away. The axon acts like a wire transmitting a signal over a long distance. Think of turning on a light using a light switch: often, the switch is many feet away from the actual lamp. The switch and the light are connected by a wire, just as a neuron's cell body and axon terminal are connected by the axon. The **myelin sheath** is a fatty layer around the axon that helps the signal travel quickly from one end of the axon to the other. This is like insulation on a wire: it stops the signal from leaking out before it reaches its destination, the axon terminal.

> **Dendrites**: thin branches on the neuron that receive messages from other neurons and send them on to the cell body
>
> **Cell Body**: contains the nucleus and organelles and is the largest compartment of the neuron. Messages from other neurons are processed here and sent to the axon.
>
> **Axon**: long, thin structure that extends from the cell body and carries messages from the cell body to the axon terminal
>
> **Axon Terminal**: the end of the axon, where messages are sent to other neurons
>
> **Myelin Sheath**: fatty tissue that wraps and insulates the axon to make signals travel faster

IB. Types of Neurons

There are many types of neurons, and they differ in several ways:

1. *Shape and appearance.*
Some neurons have very elaborate dendrites (called dendritic trees). Neurons with particularly large dendritic trees can be found in a part of the brain called the cerebellum, where they help you coordinate your movements. Other neurons have very long axons. One example is motor neurons in the spinal cord. These axons must extend from the cell body in the spinal cord all the way to their target muscles and can be over a foot long!

2. *The chemical messengers or **neurotransmitters** a neuron produces.*
Each kind of neuron sends out a different type of message to other neurons. These messages take the form of "chemical messengers,"

> **Neurotransmitter**: the chemical messenger that acts as a signal from one neuron to another

known to neuroscientists as neurotransmitters. A neurotransmitter is a small molecule released by one neuron and received by another neuron. There are many kinds of neurotransmitters that have different functions. You will learn more about what neurotransmitters do in Chapter 3. You will also learn more about specific kinds of neurotransmitters in the chapters on stress (Chapter 7) and addiction (Chapter 9).

3. *The way a neuron reacts to neurotransmitters.*
Just as there are many different kinds of messages a neuron can send, there are many ways the receiving neuron can interpret that message. Imagine two friends that have plans to get dinner together. If one friend cancels the dinner date, the other friend could react in several ways. An angry person might yell at their friend for breaking their plans, while an easygoing person might think it was no big deal. Just like different types of people can have opposite responses to the same event, neurons can have a variety of reactions to the same type of neurotransmitter message. You will learn more about how neurons interpret neurotransmitters in Chapter 3.

IC. Glia: Supporting Cells in the Brain

Besides neurons, there are several other kinds of cells that make up the nervous system. These are the "support cells" of the brain and spinal cord, and they are called glia. Glia are very important to the health and function of neurons and help them do their job well.

Here are some types of glia:

Astrocyte: the kind of glial cell that provides nutrients to neurons and regulates the environment around neurons

1. **Astrocytes** give neurons the nutrients they need, help regulate the environment around neurons, and help neurons communicate effectively with one another. Neurons need a lot of help: for every neuron, there can be as many as 10 astrocytes!

Oligodendrocyte/Schwann Cell: the kind of glial cells that provides myelin sheaths for axons. Oligodendrocytes do this in the central nervous system, while Schwann cells do this in the peripheral nervous system.

2. **Oligodendrocytes** and **Schwann cells** produce the myelin sheath, the fatty tissue that wraps and insulates axons. Oligodendrocytes do this for neurons in the central nervous system (remember that this includes the brain and spinal cord). Schwann cells make the myelin in the peripheral nervous system (remember that this includes the nerves outside the brain and spinal cord).

Microglia: the kind of glial cell that cleans up neurons' waste

3. **Microglia** are like the janitors of the brain, cleaning up after the mess that neurons make and clearing away dying cells.

To see what each of these types of glia look like, see Figure 2-2.

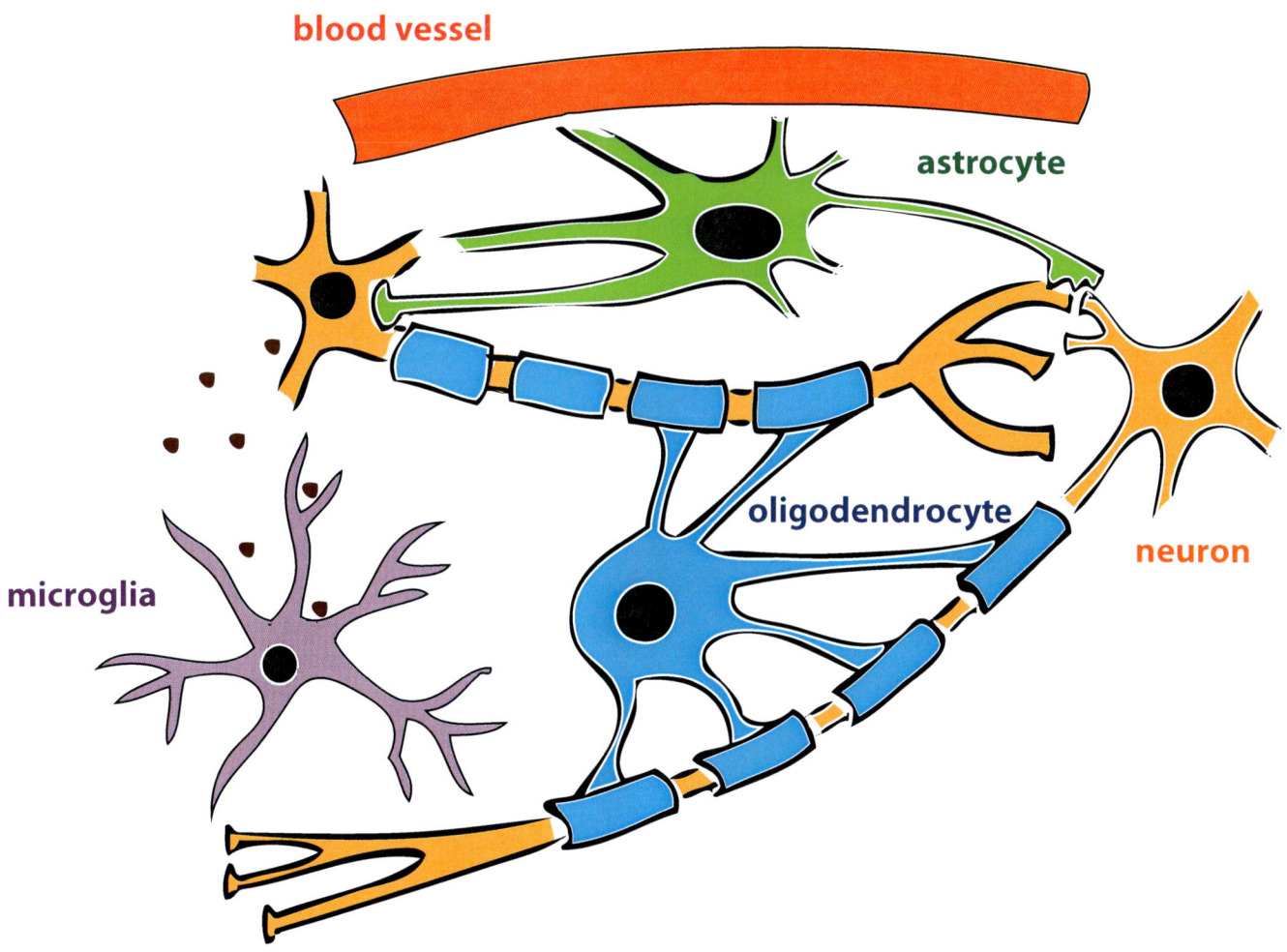

Figure 2-2: Types of Glia

II. How do Neurons Communicate?

You've now learned about the different parts of a neuron, and that neurons use neurotransmitter molecules to communicate with each other. But how are signals actually passed along from one neuron to another?

Neurons form connections called **synapses** with other neurons in order to communicate with one another. The synapse is actually a small gap between the axon terminal of the neuron sending the information (the **presynaptic neuron**) and one of the dendrites of the neuron receiving the information (the **postsynaptic neuron**). Neurons can also form synapses with other tissues, such as muscles. In this case, the synapse would be the space between the neuron's axon terminal and the muscle. To see what the synapse looks like, see Figure 2-3.

Synapse: space between the axon terminal of one neuron and the dendrite of the next neuron

Presynaptic Neuron: the neuron that is sending a message by releasing neurotransmitters into a synapse

Postsynaptic Neuron: the neuron that receives a message from another neuron by sensing neurotransmitters in the synapse

To understand the concept of neuronal communication, imagine a game of telephone. The "caller" (presynaptic neuron) talks into one cup, and the signal travels through a piece of string (the axon of the presynaptic neuron) until it reaches the cup of the "receiver" (postsynaptic neuron).

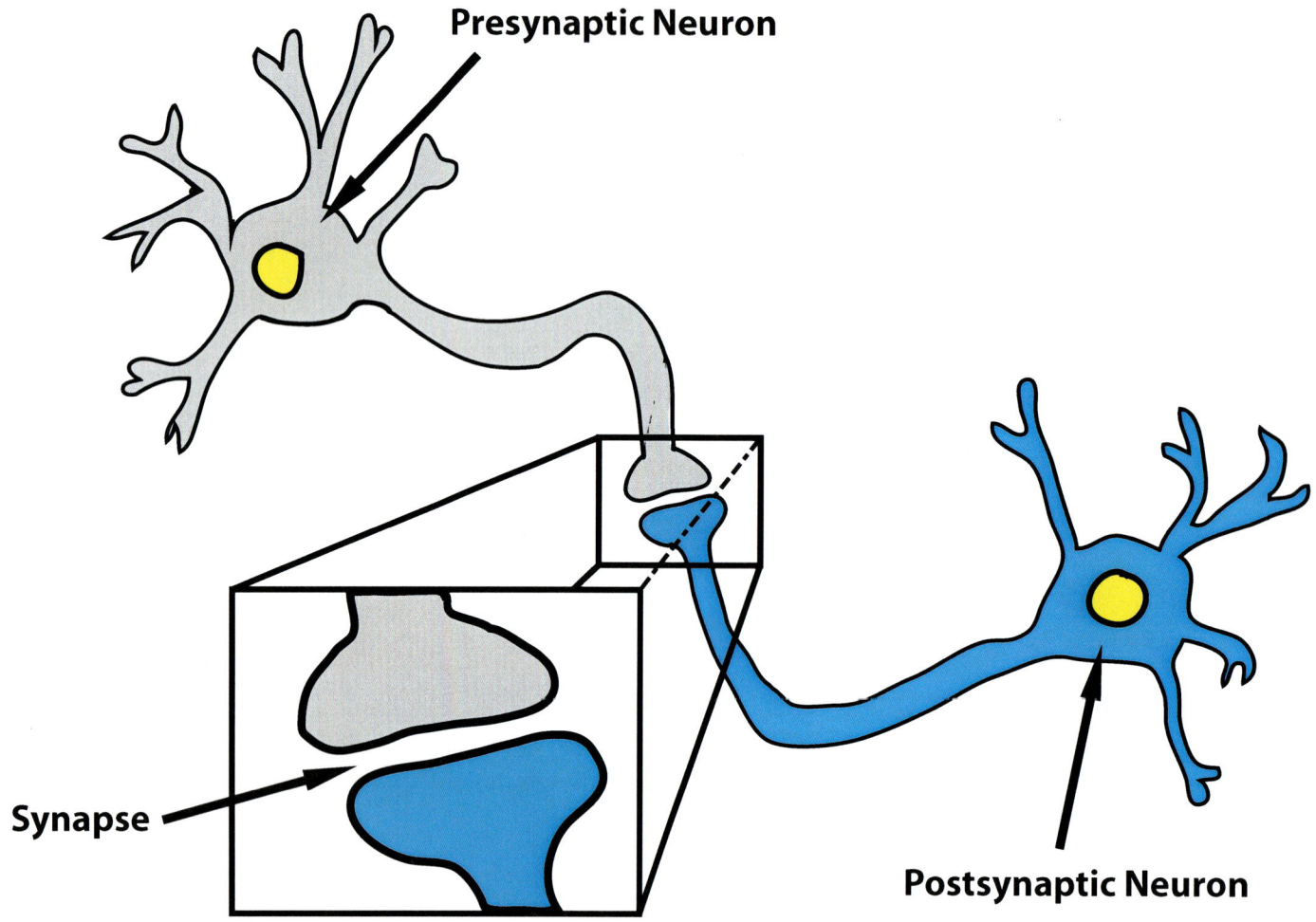

Figure 2-3: The basic structure of a synapse

Unlike this analogy, however, neuron signals can only travel in one direction: from the presynaptic neuron to the postsynaptic neuron. This is because the axon terminal of the presynaptic cell is built specifically so it can release neurotransmitter into the synapse, and the dendrites of the postsynaptic cell are built so that they can sense the neurotransmitter. Once a postsynaptic cell receives the signal, the information continues to flow in only one direction:

Dendrites ➡ **Cell Body** ➡ **Axon** ➡ **Axon Terminal**

This is how messages travel from one neuron to the next (and to the next, and to the next...).

Neurons can send messages in two different ways: chemical signals and electrical signals. Chemical signals are the neurotransmitter molecules in the synapse (which you read about in Section IB of this chapter). Once the postsynaptic neuron senses a neurotransmitter, the neuron translates that chemical signal into an electrical signal. The electricity travels through that neuron from the dendrites all the way to its axon terminal, where the signal changes back into a chemical one (when the neuron releases neurotransmitters). In other words, chemical signals travel *between* neurons, while electrical signals travel *within* neurons. You will learn more about these two types of communication in the next chapter.

Study Questions

1. Draw and label the parts of a neuron.

2. Describe the flow of information within one neuron and from one neuron to another. Which part is a chemical signal? Which part is an electrical signal?

Challenge Question

How do you think the brain would function if there were no glia? What would be the same, and what would be different?

Brain Byte

Remember that your neurons make you who you are! You have experiences because neurons are processing information about your world. The structure of a neuron is directly related to its function. For example, there are two types of neurons that carry pain information: ones that have a myelin sheath and ones that don't. When you cut yourself, you immediately get a stinging pain that is transmitted very quickly through myelinated axons. After that, you feel a throbbing pain that lasts a lot longer and doesn't feel as sharp. Axons that are *not* myelinated carry this message of throbbing pain, so the signal is transmitted much more slowly.

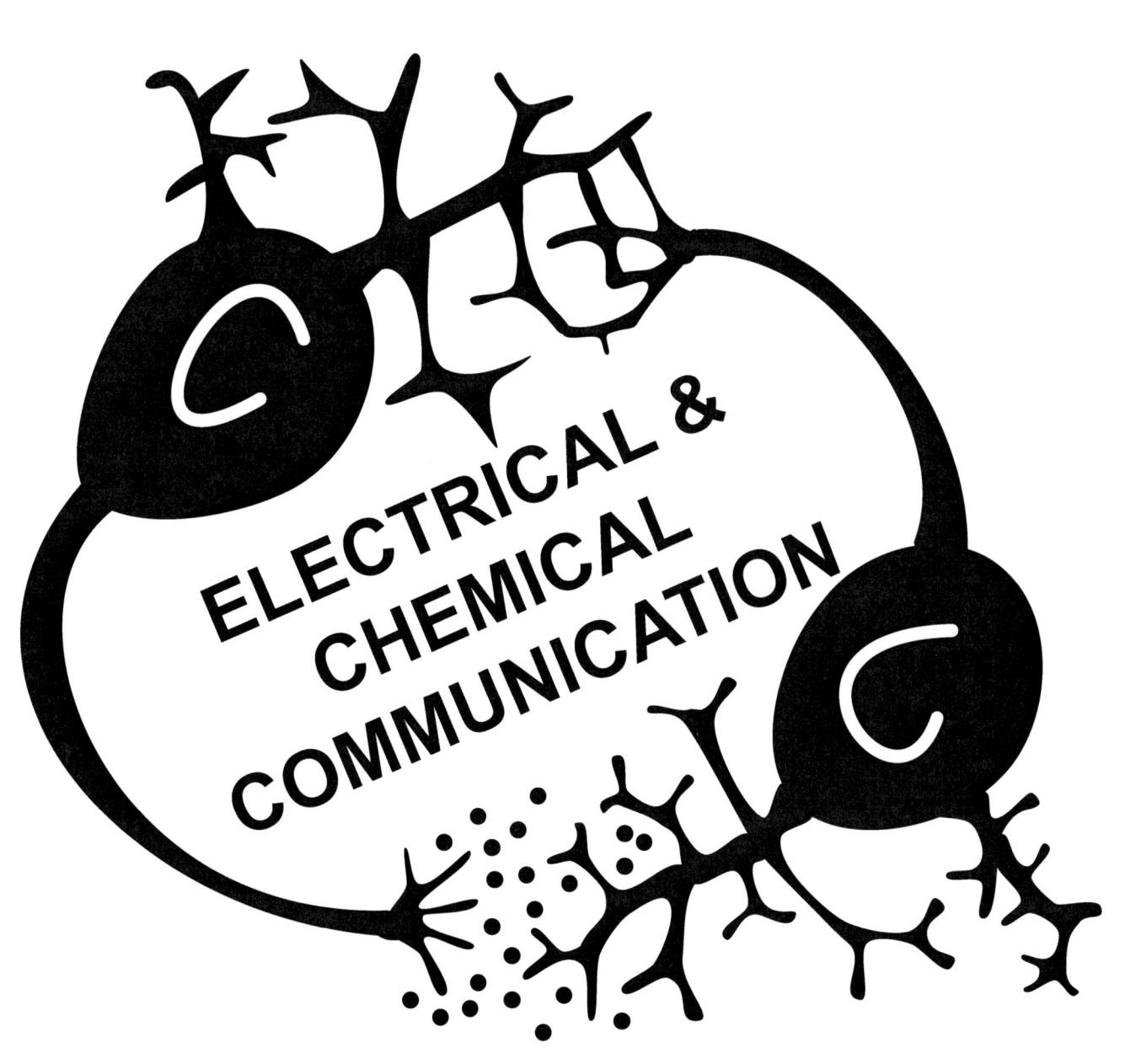

Chapter 3: Neuronal Communication

I. The Electrical Component of Neuron Communication

You've learned that there are many different types of neurons and that these neurons communicate with each other two ways: electrically and chemically. In this chapter, you will learn about these two methods of communication in more detail. When you think about electricity, what comes to mind? Lightning bolts? Charging your cell phone?

Electricity is actually the movement of particles called **ions** and is an important component of communication within a neuron. Ions are atoms that carry a charge which can be positive ("+") or negative ("-"). Positive charges are attracted to negative charges, and negative charges are attracted to positive charges. This is why there is a saying that "opposites attract!" When ions are attracted to each other's charge, we say there is an **electrical force** making them move towards each other.

To understand how neurons use electricity to send signals, we must first understand how charges (the "+" or "-" on an ion) can build up across a surface. Have you ever walked across a carpet in your socks and then felt a shock from touching a doorknob or even delivered a shock to a friend? This happens because your body has built up electrical charge. This charge then transfers from you to the doorknob in the form of electrical energy (the shock). Just like you can send a signal by shocking someone else, neurons in the brain use electrical energy to communicate.

How do neurons build up this charge? Neurons can control which ions are inside of them and which ions are outside. In general, there are more negative charges inside the neuron and more positive charges in the environment around it. Because "opposites attract," the positive charges outside the cell want to be closer to the negative charges inside the cell.

> **Ion**: a small charged particle that can pass into or out of the neuron through specialized channels

> **Electrical Force**: the attraction ions feel for ions of the opposite charge

Since the positive charges can't get to the negative charges inside the neuron, they line up along the outside surface of the cell. This surface, which separates the inside of the neuron from the outside environment, is called a **membrane**. There is a difference in charge across the membrane, meaning that one side of a membrane has more positive ions and the other side has more negative ions. We call this difference in charge the **membrane potential**. When the inside of the cell is more negative than the outside, we say the membrane potential is negative. When the inside of the cell is more positive than the outside, the membrane potential is positive. Scientists often refer to membrane potential as the membrane's voltage.

Membrane: a fatty layer that separates the inside of a neuron from the outside environment

Membrane Potential: the difference in electrical charge between the neuron's inside and the outside environment

Three important ions are involved in setting up this difference in charge. These ions are sodium (Na+), potassium (K+), and chloride (Cl-). What difference do you notice between these ions? Sodium and potassium are positively-charged ions (+), while chloride is a negatively-charged ion (-). These differently charged ions help to

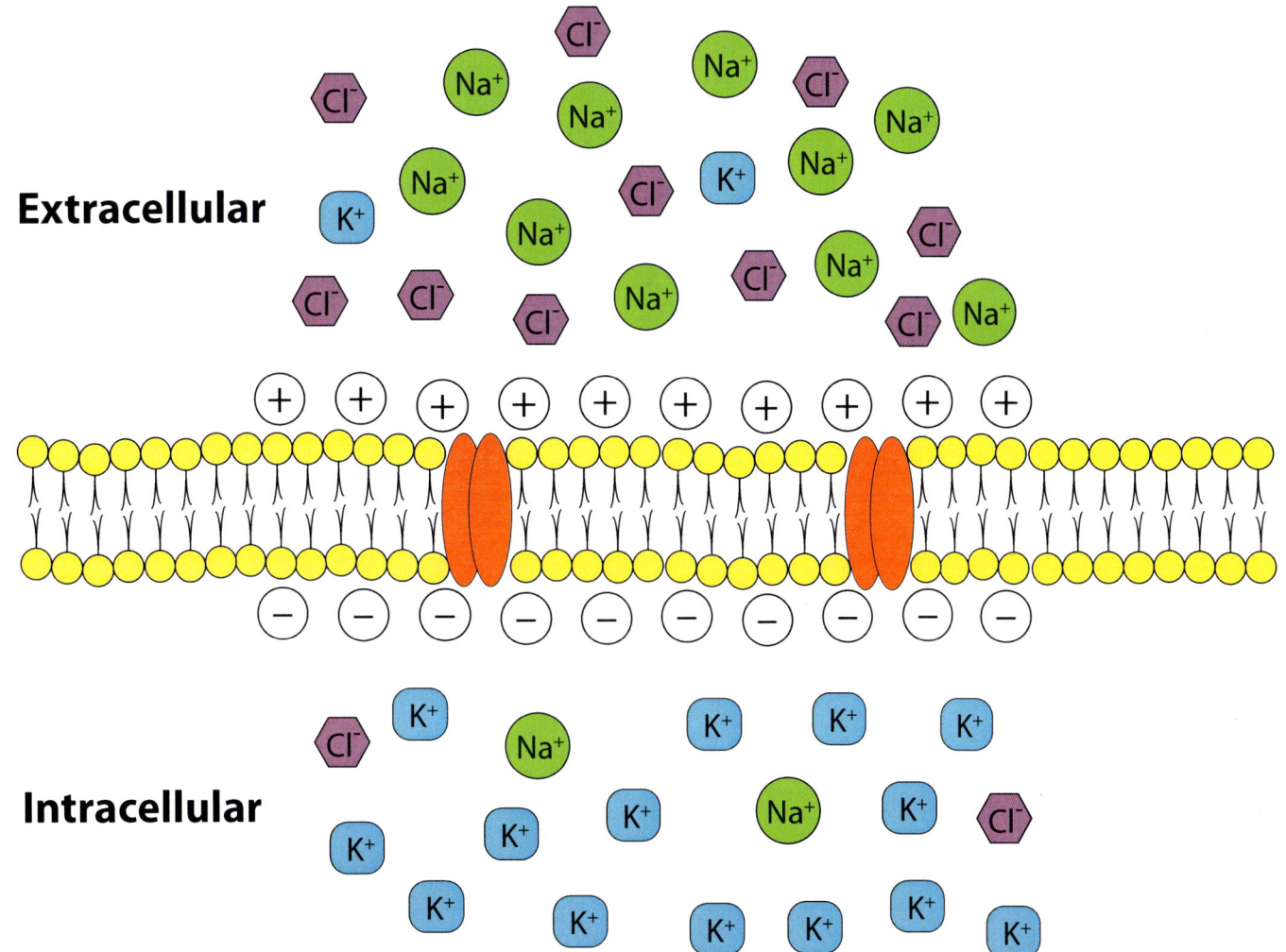

Figure 3-1: *Distribution of charged ions across the neuron cell membrane*

> **Concentration Gradient**: an uneven distribution of ions across the cell membrane. This creates a force that pushes ions from areas that have lots of that type of ion toward areas where that ion is scarce.

> **Diffusion Force**: the push ions feel to move from areas of high concentration to areas of low concentration

> **Equilibrium**: the state of an ion when it does not feel electrical force or diffusion force

> **Ion Channel**: a gate in the membrane of the neuron that ions can travel through

create the electrical component of neuronal communication.

As you can see in Figure 3-1, the neuron keeps certain kinds of ions out and makes sure to keep other kinds of ions in. This uneven distribution of ions across the membrane creates a **concentration gradient**. This means there are more ions of a certain type on one side of the membrane than the other. When this happens, we say that the side with more of one type of ion has a "higher concentration" of that ion. The word "gradient" tells you that the concentration of that ion is different on the two sides of the membrane.

Figure 3-1 shows how ions are distributed across the membrane unevenly. Which ion is more concentrated inside the cell than it is outside the cell? There are more potassium (K^+) ions inside the cell than outside, and more sodium (Na^+) and chloride (Cl^-) ions outside the cell than there are inside. Therefore, K^+ has a higher concentration inside the cell, while Na^+ and Cl^- are more concentrated outside the cell.

Why are ion concentration gradients important? Remember that ions feel an electrical force that pulls them towards other ions with the opposite charge. The concentration gradient causes ions to feel a second force, the **diffusion force**, which makes them want to go from areas of high concentration to areas of low concentration. Think of people in an elevator. When there are many people in an elevator, they will spread apart as much as they can, so there is no area that has a higher concentration of people than any other. This is the same with ions: they will try to spread out as much as they can instead of remaining concentrated in one area.

At any given time, the ions will have both electrical and diffusion forces acting on them trying to make them move one way or another. Sometimes the two forces will both push the ions to move in a particular direction, and sometimes the forces will oppose each other. The way the ions end up moving as a result causes them to reach a state of **equilibrium**, when the ions have responded to all the forces acting on them.

Now you know about the separate electrical and concentration differences across a cell membrane. There is one more piece of information that is important for understanding how neurons communicate. These important structures are **ion channels**, "gates" in the cell membrane that allow ions to pass through. They are colored orange in Figure 3-1. Ion channels allow only certain ions to travel through the membrane. There are specific ion channels only for

Na+ ions and specific ion channels only for K+.

II. The Action Potential

Now it is finally time to learn what method neurons use to communicate: the **action potential**. An action potential is a brief change or "spike" in membrane potential. You can see a graph of what that spike looks like in Figure 3-2.

> **Action Potential**: an electrical message that flows from the cell body down the axon to the axon terminal, eventually causing the release of neurotransmitters into the synapse

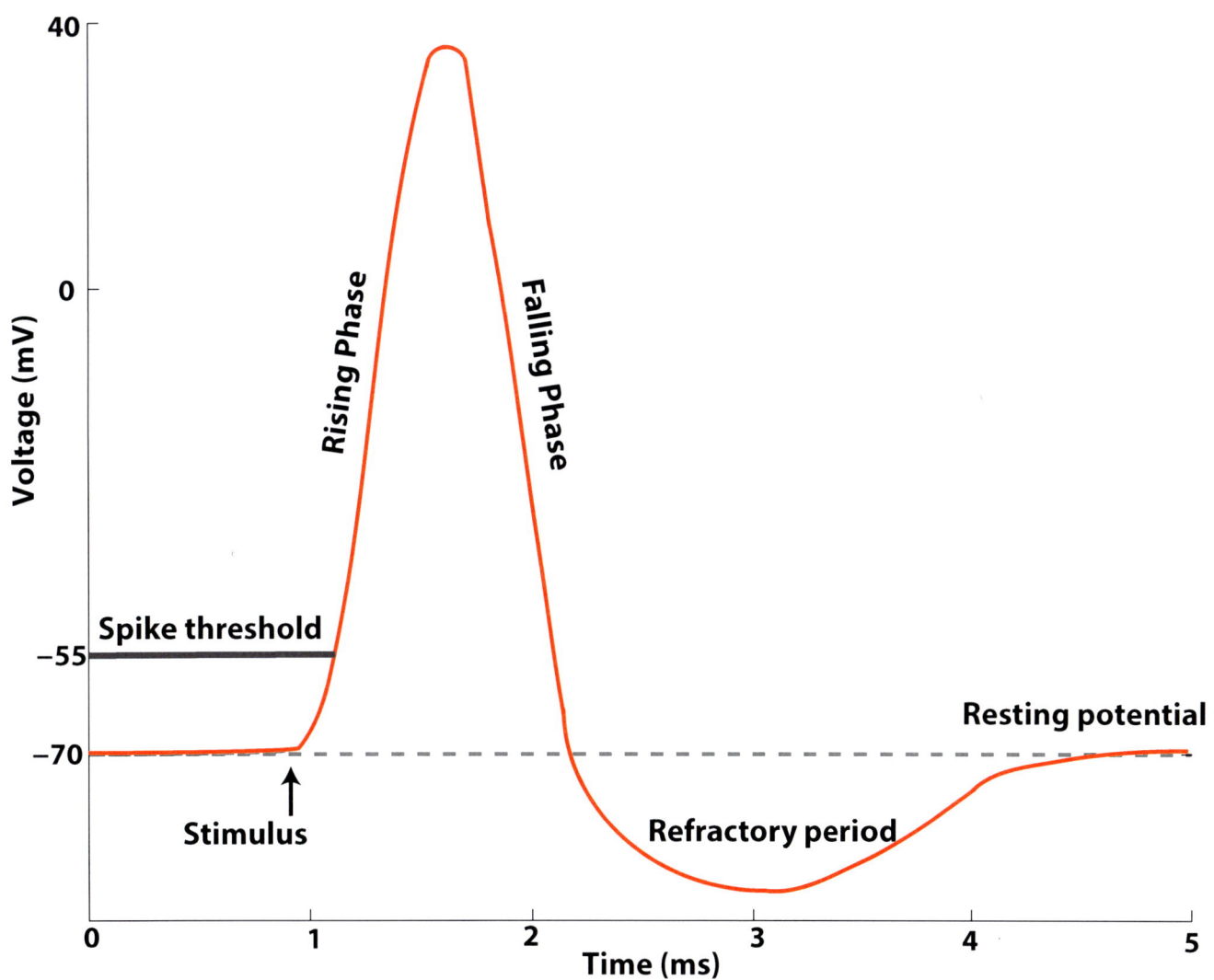

Figure 3-2: Diagram of the action potential

An action potential occurs when the inside of the membrane becomes more and more positively charged until it reaches a certain **threshold** voltage. Remember that we said the membrane potential of a neuron is typically negative (around -70 millivolts, or mV). This is because there are more positive ions outside the membrane compared to inside the cell. Thus, the inside of the cell is a negatively charged environment.

When a neuron is at rest, most of the ion channels are closed. This helps the neuron maintain its negative membrane potential. We refer to this negatively charged state (-70 mV) as the **resting membrane potential**, because this is the voltage of the cell when it is not firing action potentials. However, when neurons communicate, the membrane potential goes through several changes. You can see a diagram of these changes in Figure 3-2.

IIA. Stages of the Action Potential

1. **DEPOLARIZATION**: When the dendrites of the postsynaptic neuron receive messages from the presynaptic neuron, Na^+ ion channels open. This event allows Na^+ ions to obey the electrical and diffusion forces that drive these ions into the neuron. Remember that Na^+ is a positively charged ion and is attracted to the negative resting potential of the neuron. Furthermore, there are more Na^+ ions outside the membrane compared to inside, so Na^+ ions would like to flow inward! This increase in voltage causes the cell to be more positive and is called **depolarization**.

2. **THRESHOLD**: As more Na^+ ions flow across the membrane and make their way to the cell body, the neuron becomes depolarized (meaning its membrane potential becomes more positive). If enough Na^+ ions flow in, the neuron reaches a threshold voltage and an action potential is fired. What does this mean, exactly? To understand this, let's think about taking a picture with a camera. When you press firmly on the shutter release button, the camera shoots a photo. However, if you only lightly press on the shutter release button, the camera doesn't shoot a photo. You have to press firmly enough to trigger the shutter and once this happens, you always take a photo. The trigger in this case is similar to the threshold in a neuron. Once the cell reaches that threshold, it will fire an action potential. But, if it doesn't quite reach the threshold, it will not. This is referred to as the "all or none" action potential response.

> **Threshold**: a specific voltage that the membrane must reach before a neuron can fire an action potential

> **Resting Membrane Potential**: the negative voltage of a neuron (about -70 mV) when it is not firing an action potential

> **Depolarization**: an increase in membrane potential above resting levels

Once triggered, the action potential is the same size, every time. A neuron cannot fire half an action potential, just as your camera cannot take half a picture!

As you can see in Figure 3-2, during an action potential, the cell's membrane potential quickly rises. This positive shift in membrane potential is called the "rising phase."

3. **REPOLARIZATION:** In the process of depolarization, K^+ channels in the neuron's membrane also open, although much more slowly. By the time they open, the membrane potential is already very positive due to the massive number of Na^+ ions rushing inside the cell. Because of this positive membrane potential K^+ ions would rather flow out of the cell than in (remember, like charges repel!). K^+ is also more concentrated inside the cell. Therefore diffusion force also acts on K^+, making it want to flow out.

When K^+ ions leave the neuron, the membrane potential decreases or **repolarizes**, and this is called the "falling phase." The cell finally goes back to its resting membrane potential, completing the action potential.

> **Repolarization**: a decrease in the membrane voltage that causes a return to the resting membrane potential

IIB. The Action Potential's Journey through a Neuron

Action potentials are "traveling signals." The first action potential happens at a tiny segment of the neuron just where the cell body meets the axon. This first action potential triggers another action potential in the very next segment of membrane. In this way, the action potential moves down the axon away from the cell body and toward the axon terminal.

The action potential can travel down the axon quickly or slowly, depending on whether or not the axon is myelinated. Action potentials move faster down myelinated axons than they do down unmyelinated axons (recall that myelin comes from glial support cells). This is because there are fewer ion channels in myelinated areas of the axon, which means that there are fewer opportunities for ions to either leave or enter the neuron.

One way to think about this is to imagine a hose that has a lot of holes. Water will tend to leak though the holes and take a long time getting to the end of the hose. If you then duct tape large segments of the hose (so that there are fewer holes), the water would leak out less and arrive at the end of the hose more quickly. Myelinated axons act more like the duct-taped hose, and unmyelinated axons act more like the hose full of holes.

Up to this point, we have mainly talked about electrical neuronal communication. However, at the axon terminal, these electrical signals are turned into chemical signals in the form of neurotransmitters (recall from Chapter 2 that neurotransmitters are the chemical messages neurons send one another). Neurotransmitters will then travel across the synapse to the dendrites of the next postsynaptic neuron. This is the chemical component of neuronal communication.

III. Chemical Component of Neuronal Communication

What happens when the action potential reaches the end of the axon? The arrival of the action potential at the axon terminal triggers the release of neurotransmitters into the synapse. Remember that a synapse is the space between the axon terminal of one neuron and the dendrite of the next neuron, as you can see

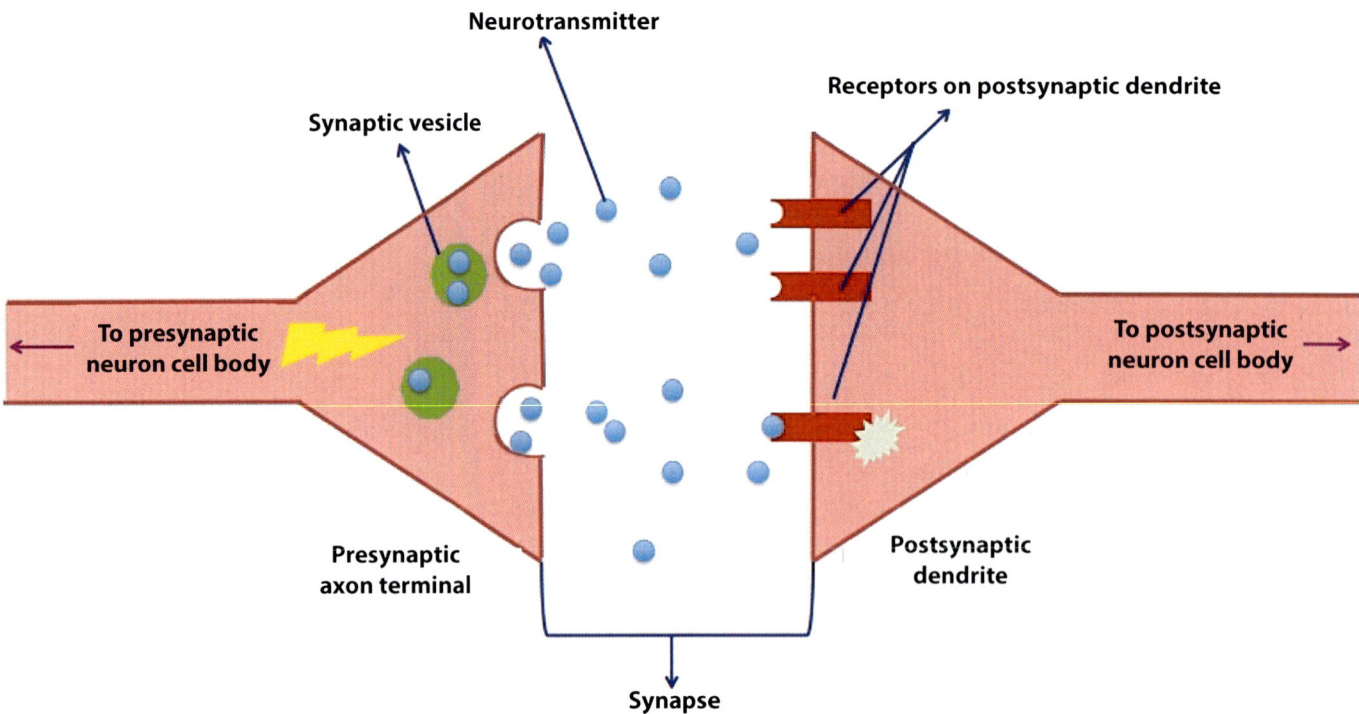

Figure 3-3: Chemical communication at the synapse

in Figure 3-3. In the synapse, neurotransmitters bind to **receptors**, which are structures specially designed to recognize specific chemical messages. These receptors are located on the dendrites of the postsynaptic neuron. Neurotransmitters bind to specific receptors similar to how a key fits into a lock.

Receptor: structure on a dendrite's surface that recognizes and binds to specific neurotransmitters

How do neurotransmitters get into the synapse? Neurotransmitters are stored in packets called synaptic vesicles. The neuron stores these packets at the very end of the axon terminal, ready to be released from the presynaptic cell.

Once a neurotransmitter binds to a receptor, the interaction can trigger many types of responses. Some neurotransmitters, like glutamate, are excitatory – they cause the membrane voltage to become more positive. This type of neurotransmitter may cause a neuron to fire action potentials more often. Other neurotransmitters, like GABA, are inhibitory – they cause the membrane voltage to become more negative. This causes neurons to fire action potentials less often.

Many neurotransmitters help control processes in our daily lives. For example, acetylcholine helps regulate learning and memory, while dopamine is important for movement control. You will learn more about how specific neurotransmitters are important for processes like sleep and drug addiction later in this book.

Study Questions

1. Describe the events that must happen in order for an action potential to fire.

2. What can affect the speed of an action potential?

3. Draw an outline of a neuron with dendrites, cell body, axon, and axon terminal. In which direction does the action potential go? Draw arrows on the neuron to show your answer.

Challenge Question

Think back to the camera analogy of the "all or none" response of an action potential. Some cameras may require a more firm press than others. How could this be similar to a neuronal action potential?

Brain Byte

How did we first discover action potentials and learn about neuronal communication? Neuroscientists Alan Hodgkin and Andrew Huxley performed experiments on a squid giant axon in the 1950s. This axon is used to move the squid's tail, and it runs all the way from the squid's head to its tail. The 1 millimeter diameter of the axon allows it to be seen with the naked eye. This permitted scientists to study the properties of axons even without the powerful microscopes we have today.

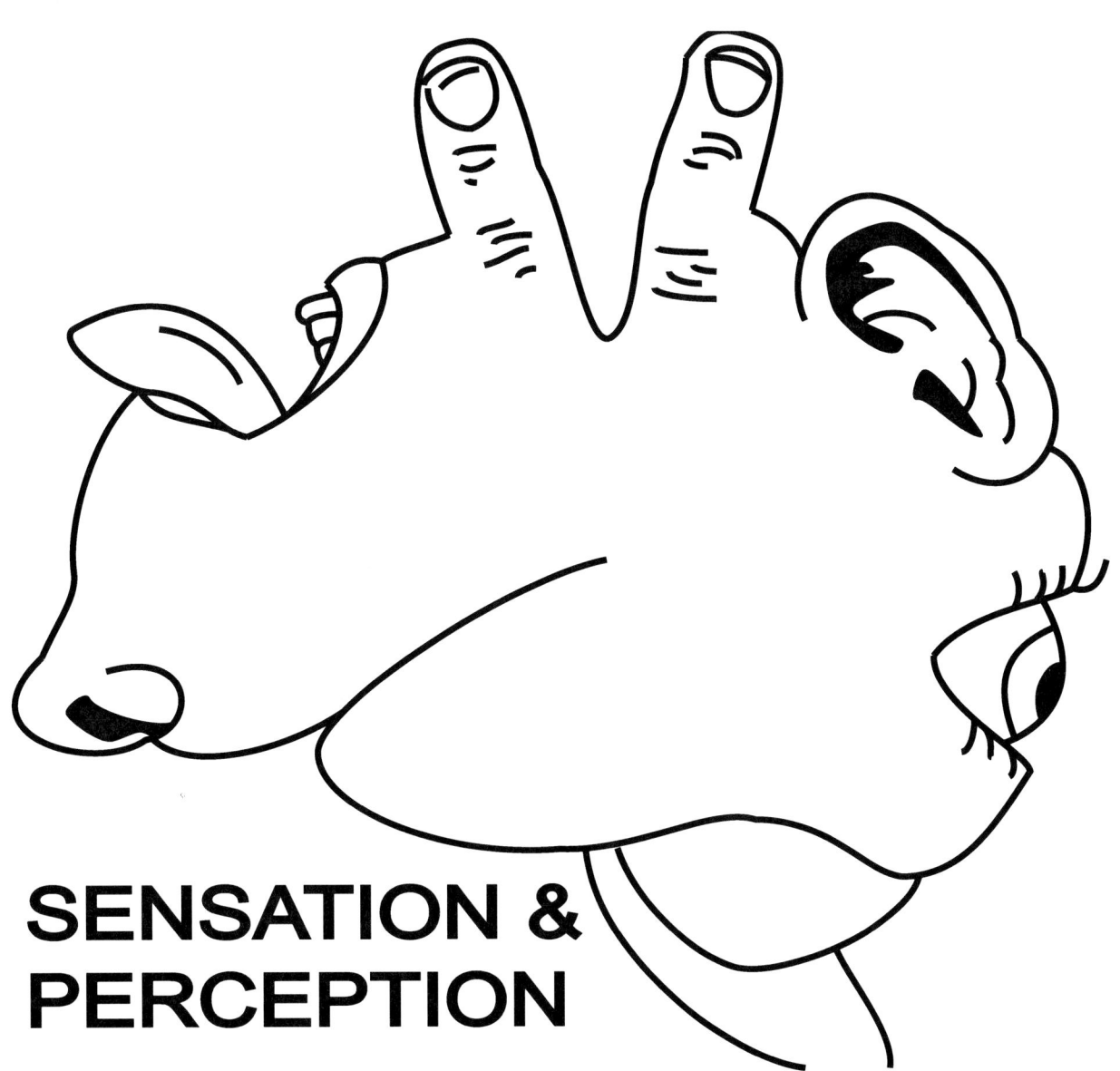

SENSATION & PERCEPTION

Chapter 4: Sensation and Perception

I. Sensation and Perception

We perceive the world through our senses. That is, we understand the world and interact with it through our senses of sight, touch, hearing, smell, and taste. To perceive something is to be consciously aware of it, and your brain is what allows this to happen. Most people assume that their senses are simply reporting what is happening in the world. However, as you will learn in this chapter, your brain takes an active role in deciding what you actually perceive. A good example of what we mean by this is visual illusions, which trick your brain into thinking it is seeing something that really is not there (for further discussion, see Section II of this chapter). We will come back to visual illusions shortly, but first it is important to understand how sensory information reaches your brain.

IA. Types of Sensation and Sensory Receptors

Sensory Receptor: a type of cell that detects basic physical properties of the world

Our five senses are sight, touch, hearing, smell, and taste. What are these senses? Why does something have a smell? Why does something make a sound? Our five senses correspond to physical properties of our environment, and we have evolved special types of **sensory receptors** that detect these properties.

SIGHT (VISION): Why do we need light in order to see? Our visual sensory receptors are designed to detect reflected light. When light shines on an object, some the light bounces off that object and back towards the eye. This lets our visual system know an object's location, color, and size.

Photoreceptor: a type of sensory receptor in the retina that detects light

Our visual system begins with the retina, which is located at the back of the eye. The retina is made of sensory receptors called **photoreceptors** that detect the light reflected off of an object. There are two types of photoreceptors in your eyes: rods and cones. We will discuss photoreceptors in detail in Section III of this chapter.

TOUCH: What can you tell about an object by using your sense of touch? You may notice how hard or soft it is, or maybe how smooth or rough it is. Our skin has sensory receptors called **mechanoreceptors** that detect pressure (hard or soft) and vibrations (smooth or rough). You may also notice the object's temperature, which tells you how fast molecules in this object are moving. Our skin can detect if something is hot or cold (the movement of molecules) using sensory receptors called **thermoreceptors**.

Finally, what if something you touch is VERY hot, so hot that it hurts your hand? An extremely hot object (for instance, a stove burner turned to "high" or a lit match) can cause the molecules of your skin to move too fast, injuring your skin. **Nocioreceptors** detect this damage and send a message of pain to your spinal cord and brain. This allows you to quickly remove your hand from the hot stove and prevent further damage.

HEARING: How does an object make a sound? Let's say you drop a heavy book on the floor and it makes a loud "bang." When the book strikes the floor, it disturbs the air around it and creates waves of pressure called sound waves. You can't see these waves, but you can hear them. You detect sound waves using mechanoreceptors in your ear called "hair cells." These "hairs" bend when sound waves enter the ear. This bending sends a signal to the brain that you are hearing a sound. Note that these mechanoreceptors (hair cells) are different from the mechanoreceptors in your skin, but they perform a similar job: both kinds of mechanoreceptors detect pressure.

SMELL AND TASTE: What do smell and taste have in common? Both of these senses involve the detection of chemical substances. When you smell something, **chemoreceptors** in your nose are detecting chemical substances in the air. When you put food into your mouth, chemoreceptors on your tongue are detecting the chemical properties of your food.

IB. From Sensory Receptor to Your Cortex

So far, you've learned that we use sensory receptors to see, touch, hear, smell, and taste things. However, sensory receptors alone are not enough! Sensory receptors have to send the information they collect to your brain, specifically the cortex, for you to actually perceive the world. To perceive something means that you are aware that you are seeing, hearing, touching, smelling or tasting that thing. As you will learn below, it is possible for the photoreceptors in your eyes to be perfectly healthy, but if your **primary visual**

Mechanoreceptor: a type of sensory receptor that detects pressure. This can be pressure or vibration in the skin or sound waves in the ear.

Thermoreceptor: a type of sensory receptor that detects temperature

Nocioreceptor: a type of sensory receptor that detects pain

Chemoreceptor: a type of sensory receptor that detects chemicals in food (taste) or in the air (smell)

Primary Visual Cortex: the first area of the cortex that processes visual information

cortex is damaged then you will not know that your eyes are detecting light. This means that though your photoreceptors are working fine, you will not be able to see.

All sensory information except smell takes the same basic pathway from sensory receptor to cerebral cortex:

Sensory Receptor ➡ Thalamus ➡ Cortex

The thalamus is a structure in the middle of your brain that is crucial for sending information to the correct part of cortex for sensory processing. For instance, the thalamus sends information about light to your visual cortex (which processes vision) and not to your auditory cortex (which processes hearing). The brain processes each of your five senses in specialized areas of the four lobes of the cortex. These areas are called **primary sensory cortex**. You can see the locations of each primary sensory cortex in Figure 4-1.

> **Primary Sensory Cortex**: an area of cortex that is the first stage of processing for a sense. Each sense has a specialized primary sensory cortex.

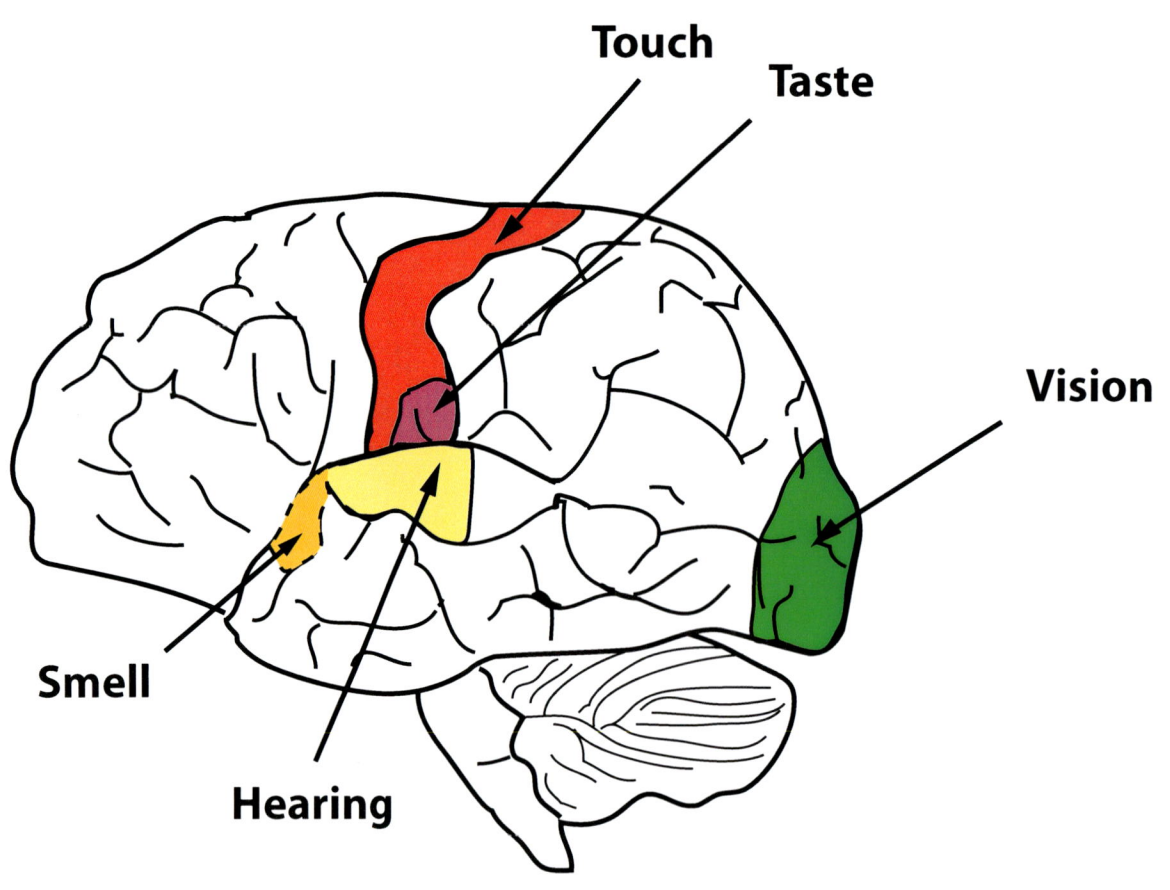

Figure 4-1: The locations of the primary sensory cortex for each sense

Visual information from photoreceptors in the retina is sent to the thalamus and then to the primary visual cortex, which is in the occipital lobe. Auditory information from hair cells is sent to the thalamus and then to the primary auditory cortex, which is in the temporal lobe. Touch information from mechanoreceptors and thermoreceptors is sent to the thalamus and then to the primary somatosensory cortex, which is in the parietal lobe. Taste information from chemoreceptors is sent to the thalamus and then to the gustatory cortex, which is in the frontal lobe. Recall that smell is unique in that the thalamus is not involved, and information is sent directly to the olfactory cortex, which is also in the frontal lobe.

II. The Cortex Produces Perception

Perception is awareness of your senses. Based on sensory information from receptors in your eyes, ears, skin, tongue, and nose, the cortex constructs your perception of the world. The cortex takes all of these different types of sensory information and combines them into a single whole that we experience as consciousness.

Perception: conscious awareness of sensory information

Another fascinating role of the cortex is to try to make predictions about what you should be experiencing from your sensory receptors. Just because you see something does not mean it is actually there. A great example of this is visual illusions.

Figure 4-2A: Shadow Checker Illusion

The illusion in Figure 4-2A is a classic.

The squares in A and B are the exact same shade of gray: A is not darker than B! Prove it to yourself by covering the other squares with your fingers. Your brain is expecting a checkerboard pattern. You see square B as lighter than it actually is (even though it is in shadow) to maintain the pattern.

Need more proof? Check out Figure 4-2B.

Figure 4-2B: Shadow Checker Illusion disambiguation

When you give your brain context (in this case, the gray bar linking squares A and B), it now knows that the checkerboard pattern is not important and that A and B are the same color.

III. Detailed Example of Sensation and Perception: Vision

So far, you've learned that we perceive the world using the following pathway:

Sensory Receptor ➡ **Thalamus** ➡ **Cortex**

Now we will take a more detailed look at one of your most complex and important senses: vision.

IIIA. How do we process light?

Most of the world we see is made up of objects that reflect light (whether from the sun or manmade lights). This reflected light shoots off in many different directions, allowing you to see the object from any angle. Therefore, when you view an object, what you see is actually a tiny amount of the total reflected light of an object; you see only the light that reflected in your direction.

Your eyeball captures this light in the following manner (refer to Figure 4-3 to follow along with this description):

Incoming light passes through the transparent cornea, then through the pupil (the hole in the middle of the opaque iris) and then through a lens (which bends the light). This helps focus the

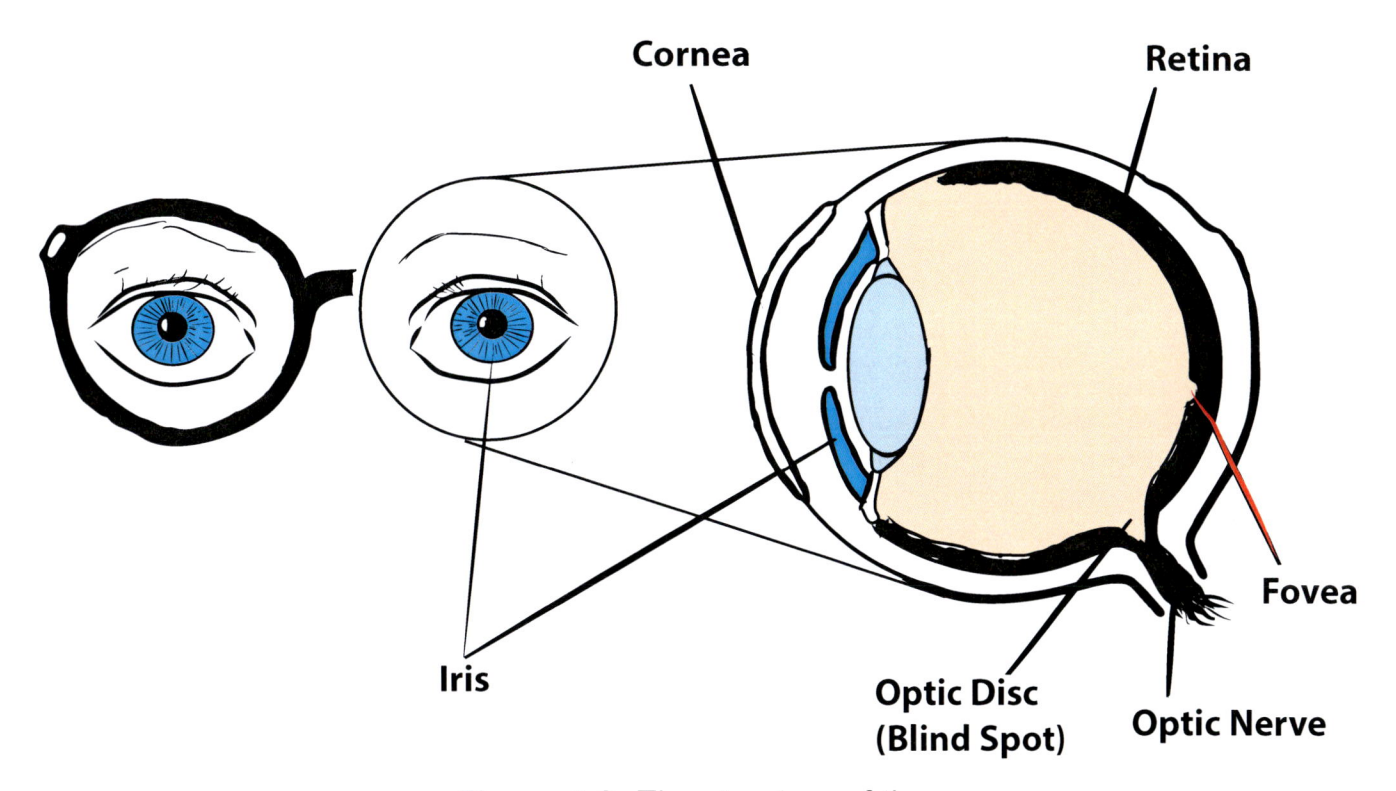

Figure 4-3: The structure of the eye

> **Retina**: the part of the eye that contains the cells that capture and process incoming light

> **Rod**: a type of photoreceptor used for seeing during the night when light levels are low

> **Cone**: a type of photoreceptor used for seeing during daylight. Cones detect an object's color.

image of the object on the back of the **retina**. The lens can change shape to bend light either more or less, allowing you to focus on near or distant objects. The pupil can also enlarge or shrink to let more or less light in. When light reaches the retina, it passes through the transparent outer layer of the retina before it is absorbed by photoreceptor cells. The objects you can see become an image on the back of the retina, similar to how a projector casts an image on a blank screen.

These photoreceptor cells come in two major classes:

> **Rods** – for nighttime vision (moonlight), and
> **Cones** – for daytime vision

There are 3 types of cones: blue, green, and red. These cones are not colored themselves, but each is named for which color of light they absorb the best. Blue cones absorb blue light very well, whereas red cones absorb blue light very poorly (though they are excellent at absorbing red light). The abilities of these cones to absorb different colors of light is due to a difference in the proteins embedded in the cone's cell membrane, not due to structural differences between the cells—all cones are the same shape.

IIIB. The Visual Pathway

> **Retinal Ganglion Cell**: a type of cell that collects signals from photoreceptors and sends these signals to the thalamus

Photoreceptors activate **retinal ganglion cells**, neurons which send their axons out the back of the eye towards the brain. Retinal ganglion cells connect to the thalamus, where most sensory information converges. The thalamus then sends this information to the primary visual cortex. The primary visual cortex processes the visual signal, then sends information about what you're seeing to areas involved in more complex visual tasks (like recognizing faces and objects). For a diagram of this pathway, see Figure 4-4. All of this communication and processing results in your conscious experience of vision.

> **Subcortical Projections**: axons that synapse onto neurons that are not part of the cortex. For example, the retina connects not only to the thalamus, but also to the brainstem. This pathway allows you to see objects without consciously knowing you can see them (as in blindsight).

You've learned that the retina sends information to the thalamus, which passes that information to the primary visual cortex. This is what allows you to be conscious of what you're seeing. However, the retina also sends information to other brain regions, including the midbrain. These pathways are called **subcortical projections**. Subcortical literally means "below the cortex," and that's exactly where these axons go.

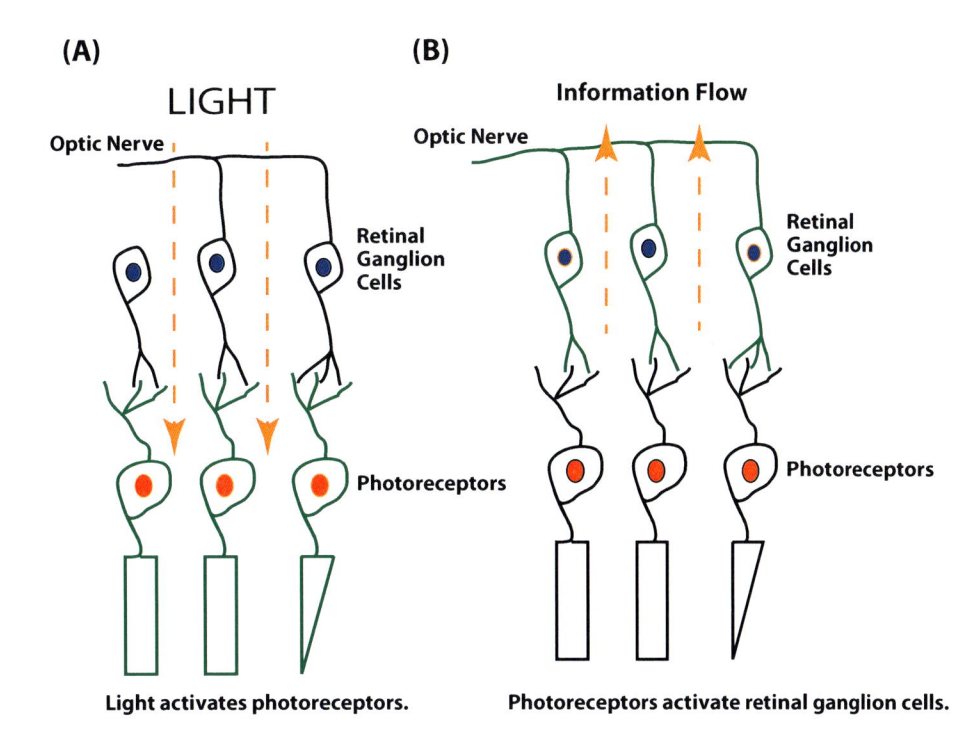

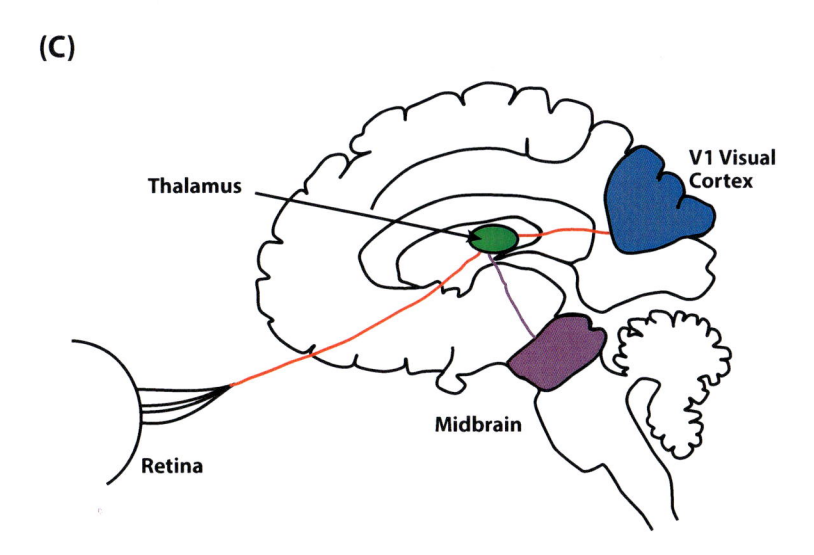

(A)

LIGHT

- Optic Nerve
- Retinal Ganglion Cells
- Photoreceptors

Light activates photoreceptors.

(B)

Information Flow

- Optic Nerve
- Retinal Ganglion Cells
- Photoreceptors

Photoreceptors activate retinal ganglion cells.

(C)

- Thalamus
- V1 Visual Cortex
- Midbrain
- Retina

Information flows from the retinal ganglion cells through the optic nerve to the thalamus and finally V1. Information also flows to the thalamus and then to the midbrain via subcortical connections.

Figure 4-3: The flow of visual information

The midbrain is important for subconscious visual processing. One illustration of this is **blindsight**, a medical condition in which the patient responds to visual stimuli without consciously perceiving them. For example, one patient with normal vision had a severe stroke that destroyed his entire primary visual cortex. He was legally blind after the stroke—even if you stood directly in front of him, he would report that he couldn't see you. However,

Blindsight: the ability to know the locations of objects without any conscious perception of light

he still had some visual processing intact. For example, the doctors had him walk by himself down a long hallway with many obstructions lying on the floor. He was able to walk the length of this hallway while avoiding all the objects. His remarkable ability to avoid obstacles he couldn't consciously see was explained by the fact that his subcortical projections, including the retina-thalamus-midbrain connection, were still intact.

IIIC. Receptive Fields

In the retina, photoreceptors will respond to any increase or decrease in the amount of light that hits that photoreceptor. Light bouncing off an object triggers photoreceptors to respond. Because each photoreceptor is permanently located at one specific spot on the retina, an object has to be located at a certain place in the visual scene in order to activate any particular photoreceptor. For example, if a photoreceptor is in the far left part of the retina, then an object needs to be in the far right area of the visual scene in order to activate that photoreceptor. This property is called a **receptive field**, the area of the visual scene that activates a particular photoreceptor.

Once the information about an object reaches higher levels of processing, different information about that object can activate different types of neurons. For example, some neurons in the primary visual cortex only activate when an object is moving across its receptive field in a certain direction. This is called **selectivity**. An object moving from left to right might activate a particular cell, but that same object moving from right to left will not activate this same cell.

> **Receptive Field**: the area of the visual scene to which a particular cell can respond

> **Selectivity**: when cells only respond to specific types of sensory information. For example, some cells will only fire action potentials when spots of light move to the right, and some cells will only fire when spots move to the left.

Study Questions

1. Name each type of sensation and match it with the appropriate sensory receptor or receptors.
2. Name each type of sensation and match it with the appropriate part of the cortex.
3. Draw the anatomy of the eye and then draw arrows indicating the path light takes to get to the retina. Where are the sensory receptors for vision located? What are they called?
4. We first detect visual information in the eye. Where does that information go when it leaves the eye? Be sure to include all areas, including any midpoints before the information reaches its final destination.

Challenge Question

Compare the mechanoreceptors on your fingertips to the mechanoreceptors on the back of your hand by lightly touching each. Which mechanoreceptors are more sensitive to the touch? Do you think it takes a lot of pressure (high threshold) or a little pressure (low threshold) to trigger mechanoreceptors on your finger? On the back of your hand? Why do you think this is?

Brain Byte

Researchers can now artificially switch off parts of the brain with a technique called Transcranial Magnetic Stimulation (TMS). This technique effectively shuts down an area of the brain for a short amount of time. Because TMS doesn't require any surgery, it can be used in human research. For example, if a doctor uses TMS to shut down a patient's primary visual cortex, that patient would lose the ability to see for a short amount of time. TMS can also be used on more specialized brain areas that occur later in the visual pathway. For example, by inactivating a specific area of the brain involved in seeing movement, the patient would be unable to see moving objects but would still be able to see stationary objects.

LEARNING AND MEMORY

Chapter 5: Learning and Memory

I. What is Memory?

Stages of Memory

> **Encoding**: the processing and organization of information into unstable short-term memories

Memory is the ability to encode, store, and recall information. Memory can be thought of in two stages: 1) short-term memory and 2) long-term memory. When we initially encounter new information, it enters short-term memory, which can hold a small amount of information for a short period of time. **Encoding** is the processing and organization of information in short-term memory.

> **Consolidation**: the process we use to turn unstable, short-term memories into long-term memories which can be recalled at a later time

We then must prepare these short-term, encoded memories to enter long-term memory. **Consolidation** is the process we use to stabilize, store, and translate information from short-term memory to long-term memory. Consider the example of trying to remember a phone number long enough to dial it. You might keep repeating the ten digit number over and over to yourself so that you won't forget it before you key it in to your phone. In this case, you are keeping that number in your short-term memory. However, even five minutes later, how often can you still remember all ten digits? You forget them because you did not consolidate the memory for this phone number into long term memory!

> **Retrieval**: the memory process we use to recall consolidated information

There is one more crucial step: in order to recall consolidated memories, we must use a process called **retrieval**. Retrieval is how we fetch memories out of long-term storage so that we can think, talk, or write about them.

Illustrated in Figure 5-1 are some parts of the brain important for learning and memory.

> **Declarative Memory**: memory for everyday facts and events

II. Types of Long-term Memory

48

There are two divisions of memory: 1) **declarative memory** and 2) **non-declarative memory**. Figure 5-2 is a diagram you can use to help you visualize the different categories and sub-categories of memory.

Non-declarative Memory: reflexive, automatic knowledge of how to do something

Cerebellum (non-declarative/procedural memory)

Hippocampus (declarative memory)

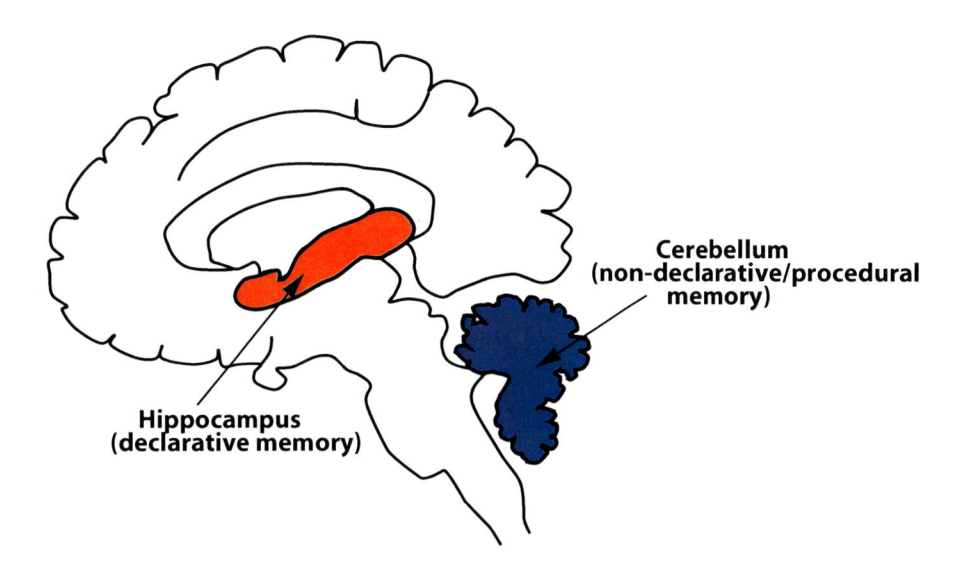

Figure 5-1: *Regions of the brain that are important for learning and memory*

IIA. Declarative Memory

Declarative memory is our memory for everyday facts and events. Declarative memory can be further sub-divided into two types: 1) **semantic memory** and 2) **episodic memory**. Semantic memory is our knowledge of words and concepts. For example, you know that Washington, DC is the capitol of the United States. You know this to be true without needing to also remember when or where you learned this fact. This is semantic memory. Can you think of other examples?

Semantic Memory: a type of declarative memory for general facts and concepts

Episodic Memory: a type of declarative memory for specific events

Episodic memory is how we remember events in space and time. In plainer language, it is how you remember the location and timing of a sequence of events from your life. When you remember what cereal you ate for breakfast, you're recalling an episodic memory. However, remembering that cereal is a breakfast food is a semantic memory. If your teacher asked to you to name a few breakfast foods, you would need to access semantic memory. If he or she asked you to make a list of what food you had for breakfast every day this week, you would need to access episodic memory.

49

Anatomy of Declarative Memory

> **Hippocampus**: a region of the brain in the middle of the temporal lobe that is important for declarative memory

The **hippocampus** (which you can see in Figure 5-1), located in the middle of the temporal lobe, is critical for encoding memories and consolidating these memories for long-term storage. How do we know this? We learned about memory from a famous patient known to by his initials, H.M. This patient had the portion of his brain that contains the hippocampus completely removed. After the surgery, he could no longer make new episodic memories!

IIB. Non-declarative Memory

Non-declarative memory, sometimes called procedural memory, is our reflexive, automatic knowledge of how to do something. It allows us to carry out common tasks without consciously thinking about them. Some examples of this type of memory are riding a bike, tying your shoes, and using a pencil to write your name. At the beginning, learning these tasks involves lots of repetition and practice. However, once we form a procedural memory for a given task, we no longer have to consciously think about how to perform that task.

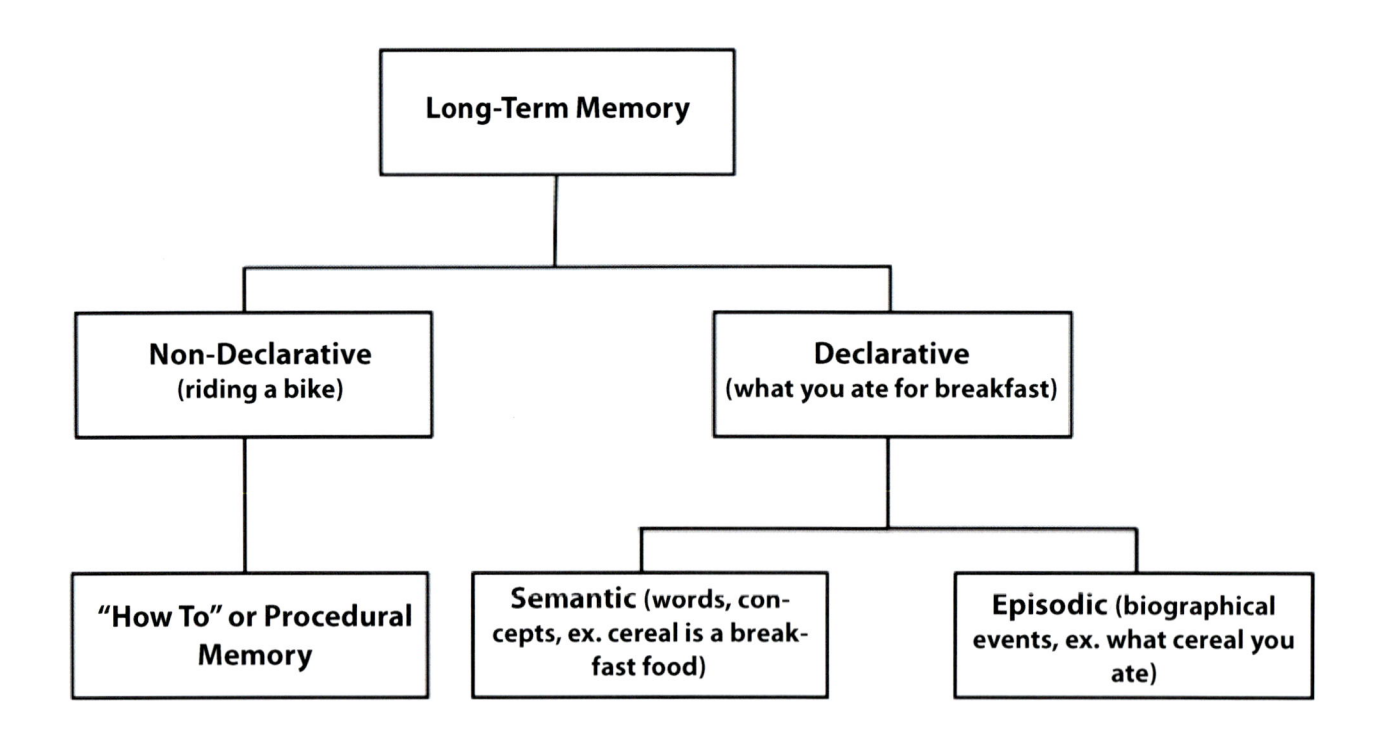

Figure 5-2: Divisions and sub-divisions of long-term memory

Once you learn how to ride a bike, you can automatically carry out the sensorimotor behaviors you need to pedal, keep your balance, and turn corners without tumbling to the ground. Your procedural memory for "how to ride a bike" allows you to automatically do all of these things without consciously giving your legs, arms, and torso specific instructions. Can you think of some other examples of procedural memory?

Anatomy of Non-declarative Memory

Non-declarative memory uses a different part of the brain from declarative memory. One important structure involved in procedural memory is the cerebellum, which is the structure at the very back and bottom of the brain (pictured in Figure 5-1). You use your cerebellum to coordinate movements in a smooth and accurate manner and to organize the timing of these movements. For example, your cerebellum helps you swing a bat at a baseball in a smooth, arcing movement at just the right time to connect with the ball.

Let's return for a moment to the patient H.M. Even though he had his hippocampus removed on both sides, he could still make new procedural memories. This showed doctors and scientists that declarative and non-declarative memories are stored in different locations in the brain.

III. How do Neurons form Memories?

Recall that neurons form synapses with other neurons in order to communicate. We actually store information in the brain by changing the strength of these connections. As information enters the brain, certain synapses get stronger (we say they are "**potentiated**") while other synapses get weaker (we say they are "**depressed**"). This is how the brain decides which information is worth storing.

This process of changing synapse strength is called **synaptic plasticity**. Typically, if two neurons fire at the same time, the synapses between these two cells will get stronger and the connection will be potentiated. Conversely, if two cells have a strong connection to each other but rarely fire together, the strength of this synapse will decrease and the connection will be depressed. An easy way to remember this is the phrase, "neurons that fire together, wire

Potentiation: the strengthening of the connection between two neurons. Potentiation occurs when two cells communicate by firing action potentials at the same time

Depression: when the connection between two neurons gets weaker. Depression occurs when two cells no longer fire at the same time.

Synaptic Plasticity: change in strength of synapses. This is how the brain stores information

together."

Figure 5-3 shows a simplified learning event between two neurons. In Figure 5-3A, the synapse between the presynaptic neuron (blue) and the post-synaptic neuron (yellow) is weak. However, if both neurons fire at the same time (illustrated by the red lightning bolts in Figure 5-3B), the synapse is potentiated. Potentiation of the synapse is illustrated in Figure 5-3C as a thicker synapse projecting from the presynaptic neuron to the post-synaptic neuron and by the greater numbers of receptors on the post-synaptic neuron. Depression of a synapse is when the strength of the synapse becomes weaker over time. This occurs if neurons no longer fire together.

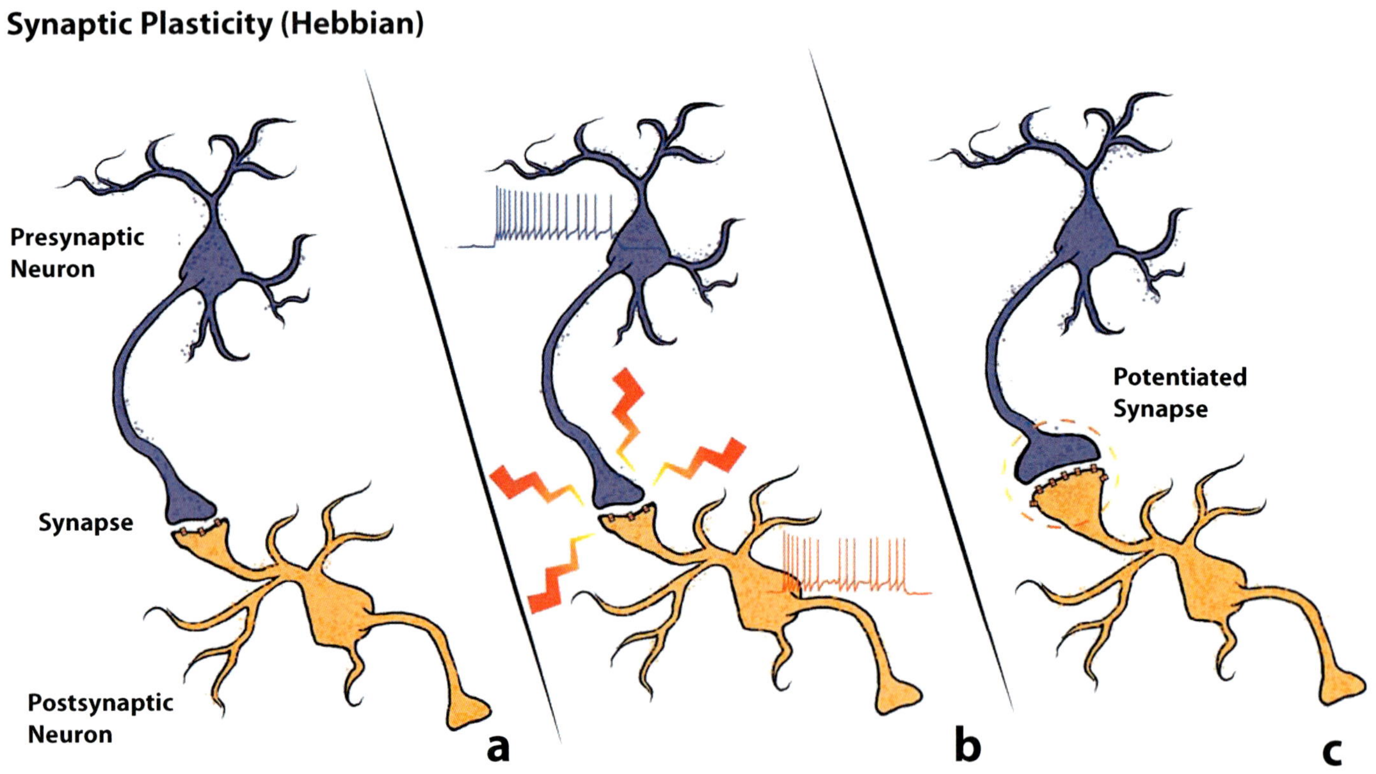

Synaptic Plasticity (Hebbian)

Figure 5-3: Hebbian learning, a type of synaptic plasticity

Study Questions

1. Which brain structure is important for declarative memory formation?
2. Come up with your own example of an episodic memory, a semantic memory, and a procedural memory.
3. How do neurons store information? (Hint: it takes more than one neuron.)

Challenge Question

Your teacher presents you with a list of words: apple, chair, book, drawer, and bottle.
(i) Your teacher asks you to recall this list immediately after. What kind of memory are you using?
(ii) Twenty minutes later, your teacher asks you to recall the same list again. What kind of memory are you using now?
(iii) Will patients with damage to the hippocampus have trouble with this memory test? Which type of memory (the answer from part i or the answer from part ii) will be more impaired? Explain your reasoning.

Brain Byte

We have learned a great deal about the function of different brain areas from patients with brain damage, like H.M. While H.M. had parts of his brain removed, scientists also study patients with much smaller injuries, which we call "brain lesions". Parts of these patients' brains no longer function because neurons in a particular area have died. A stroke (in which the brain is deprived of oxygen for a period of time) or a neurodegenerative disease (such as Alzheimer's disease) can cause brain lesions. When these patients participate in scientific experiments, we can begin to understand the function of injured brain regions based on what these patients can and cannot do.

SLEEP

Chapter 6: Sleep

I. What is Sleep?

How do you know if an organism is sleeping? A sleeping animal will move very little and assume a typical body posture (for example, humans lie down and bats hang upside-down while sleeping). Animals will react to weak stimuli, such as a soft noise, when they are awake but not when they are asleep. However, a strong stimulus, such as a loud noise, will wake an animal up. This distinguishes sleep from coma.

Electroencephalography (EEG): measurement of the electrical activity of a large number of neurons at once

The brain has unique activity patterns during sleep, which can be measured by **electroencephalography** (EEG). Action potentials from a single neuron are too small to detect from outside of the head. However, the sum of the action potentials produced by a large number (thousands to millions) of neurons is large enough to record from electrodes placed on the scalp. For a picture of this, see Figure 6-1. The signal recorded from these electrodes is the EEG.

One way to picture how an EEG works is to imagine you are in a baseball stadium where all of the fans are cheering. The sum of everyone's voices determines how loud the whole crowd sounds. If everyone starts chanting at the same time, the words of the chant will sound very loud, and in between the words it will be very quiet. Now imagine that the fans are all cheering, but they're yelling different cheers at different times. In this case, the crowd won't sound as loud overall even though just as many people are cheering.

Each of the neurons in a region of the brain is like one person in the baseball stadium, each action potential is like that one person cheering, and the EEG signal is like the sound of the whole crowd. When many neurons fire action potentials at the same time, all the electrical signals from these neurons will add up, so the EEG signal will be very large (Figure 6-1B). When neurons fire at different times, the EEG signal will be much smaller, since the electrical signals will not add up (Figure 6-1C).

During waking, the EEG signal is small, meaning that the neurons are firing action potentials at different times. As we will discuss below, the EEG can look very different during sleep.

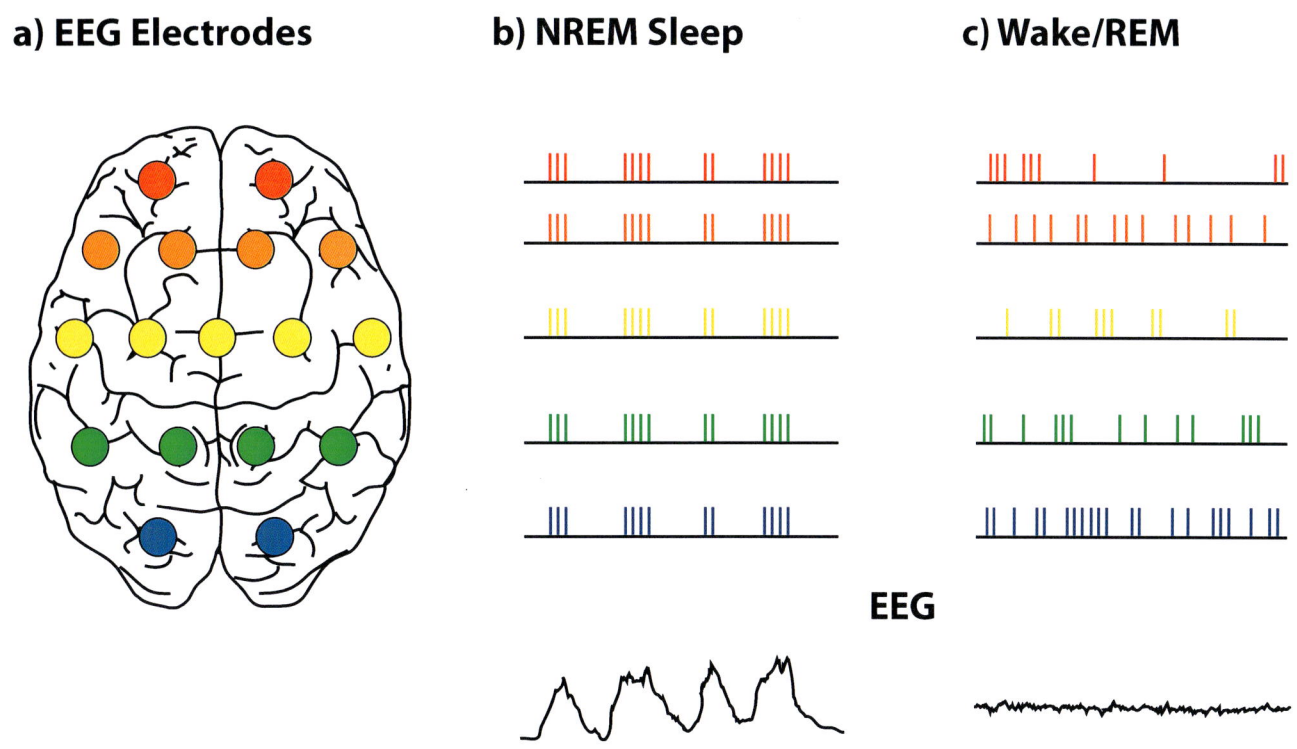

a) EEG Electrodes **b) NREM Sleep** **c) Wake/REM**

Figure 6-1: *Electroencephalography during NREM sleep, REM sleep, and waking*

II. Sleep States

During sleep, the brain alternates between two states: **non-Rapid Eye Movement** (NREM) sleep and **Rapid Eye Movement** (REM) sleep. We will discuss the differences between these states below.

Non-Rapid Eye Movement (NREM) Sleep: stage of sleep with a slow, large EEG signal

Rapid Eye Movement (REM) Sleep: stage of sleep with absence of muscle activity, a fast and small EEG signal, and rapid eye movements

NREM is the first sleep state, and it occurs just after falling asleep. It starts out as a light sleep and then becomes deeper the longer a person is asleep. EEG activity slows down and becomes larger, meaning that many neurons are firing at the same time. You can see an example of this type of EEG activity in Figure 6-2. Muscles relax, but there is still some muscle activity present, and eye movements are slow or do not occur. During NREM, dreaming can occur, but these dreams are often hard to remember after waking up. Dreams in NREM are very logical—for instance, you might dream that you are doing something normal like sitting in class taking notes or fixing yourself a sandwich for lunch.

> **Atonia**: absence of muscle activity

REM sleep only occurs after a period of NREM sleep. During REM, the EEG signal becomes fast and small, meaning that neurons are firing quickly and at different times (see Figure 6-2). This is very similar to EEG activity during waking. As the name suggests, rapid eye movements occur during REM. Muscle activity is completely blocked except for occasional muscle twitches. This is called **atonia**. This atonia causes temporary paralysis in order to prevent the sleeper from acting out dreams. You are more likely to be able to remember dreams that occur during REM sleep after you wake up. These dreams are more vivid, irrational, and emotional than NREM dreams. For instance, during REM you might dream that you are being chased by your homework through a forest of lollipops, or that the president is sitting in your kitchen feeding your sandwich to his pet rhinoceros.

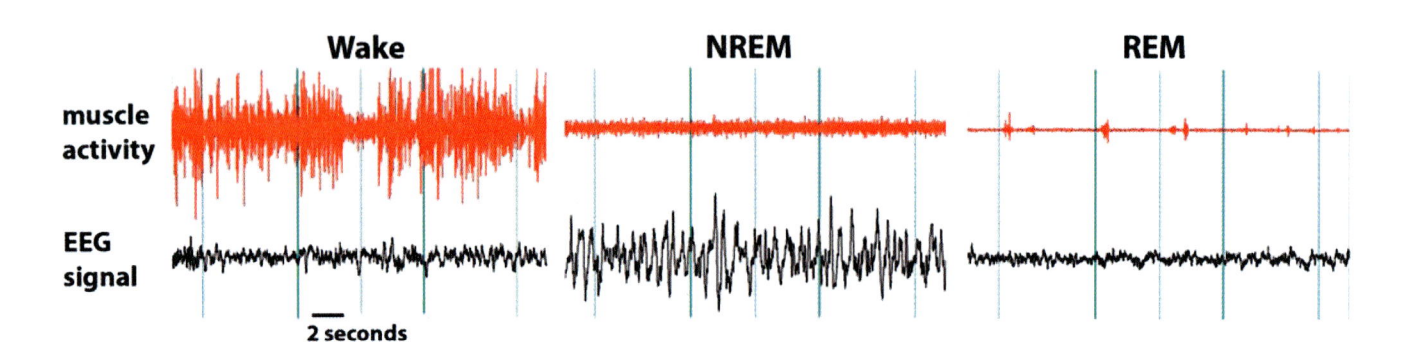

Figure 6-2: Sample EEG and muscle recordings during wake, NREM, and REM

III. How does the Brain Control Sleep?

Many different neurotransmitters in the brain affect sleep. Recall that neurotransmitters are chemicals that neurons release in order to communicate with one another. Neurotransmitters made by cells in the locus coeruleus (LC), lateral hypothalamus (LH), and forebrain are released throughout the brain to keep it awake (Figure 6-3, top). When it is time to sleep, neurons in another part of the hypothalamus called the ventrolateral preoptic area (VLPO; Figure 6-3, bottom) release different neurotransmitters that shut down the "awake" signal. This causes the brain to go to sleep. This circuit acts like a light switch that the brain can flip to turn sleep on or off. How does the brain know when it is time to flip that switch?

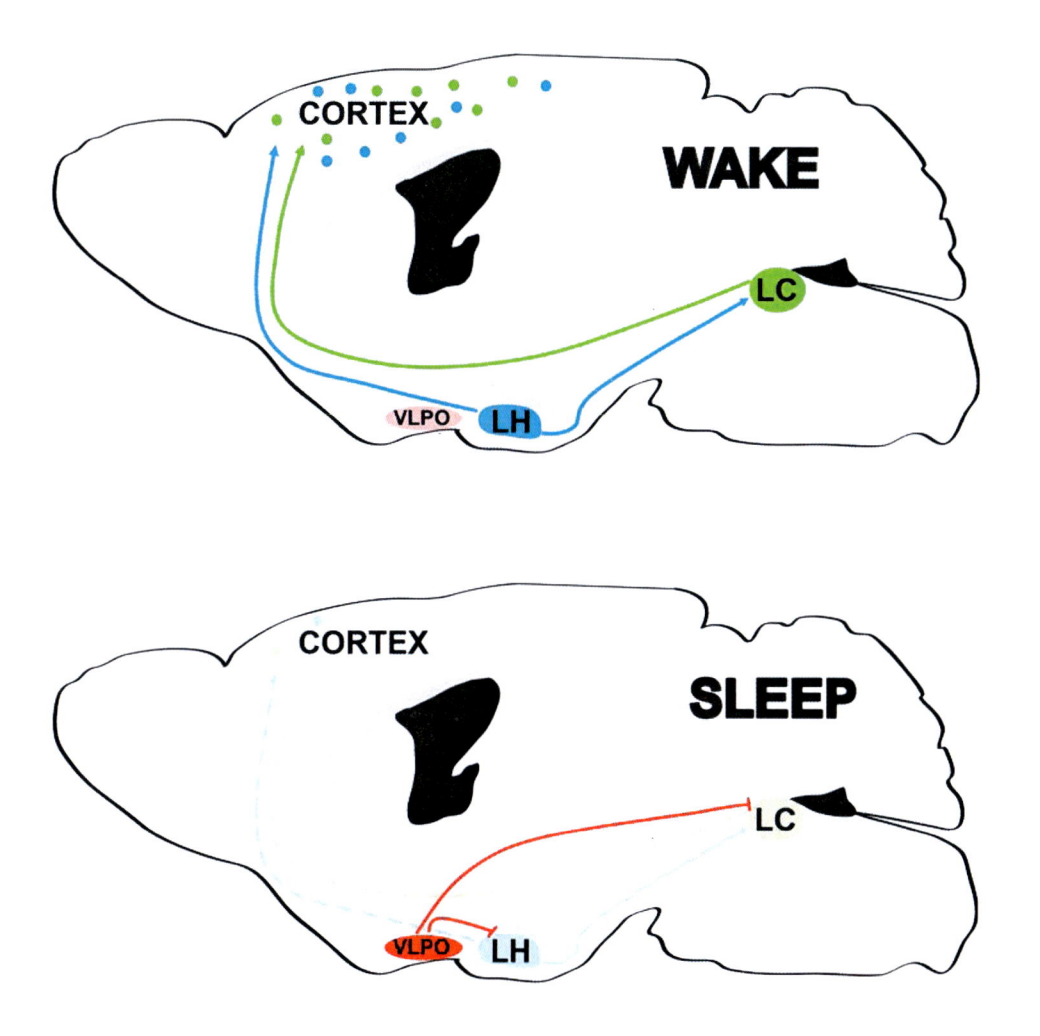

Figure 6-3: Brain regions involved in maintaining wake (top) and sleep (bottom)

There are two main factors that help determine when the brain switches sleep on or off (Figure 6-4). One is **circadian rhythm**, which is the body's internal clock. It keeps track of the time of day, which lets people stay awake during the day and wind down at night. The circadian rhythm keeps track of time regardless of surroundings, even in constant darkness. However, cues such as daylight and mealtimes help keep it accurate. The circadian clock then sends signals to the hypothalamus (Figure 6-3) to tell the brain when it is time to sleep. The importance of circadian rhythms becomes clear when you travel to another time zone. Once you leave the time zone you're used to, your internal circadian clock will no longer match up with the actual time of day. This makes it hard to fall asleep at the right time. You might know this problem as "jet lag."

Circadian Rhythm: internal clock that keeps track of the time of day

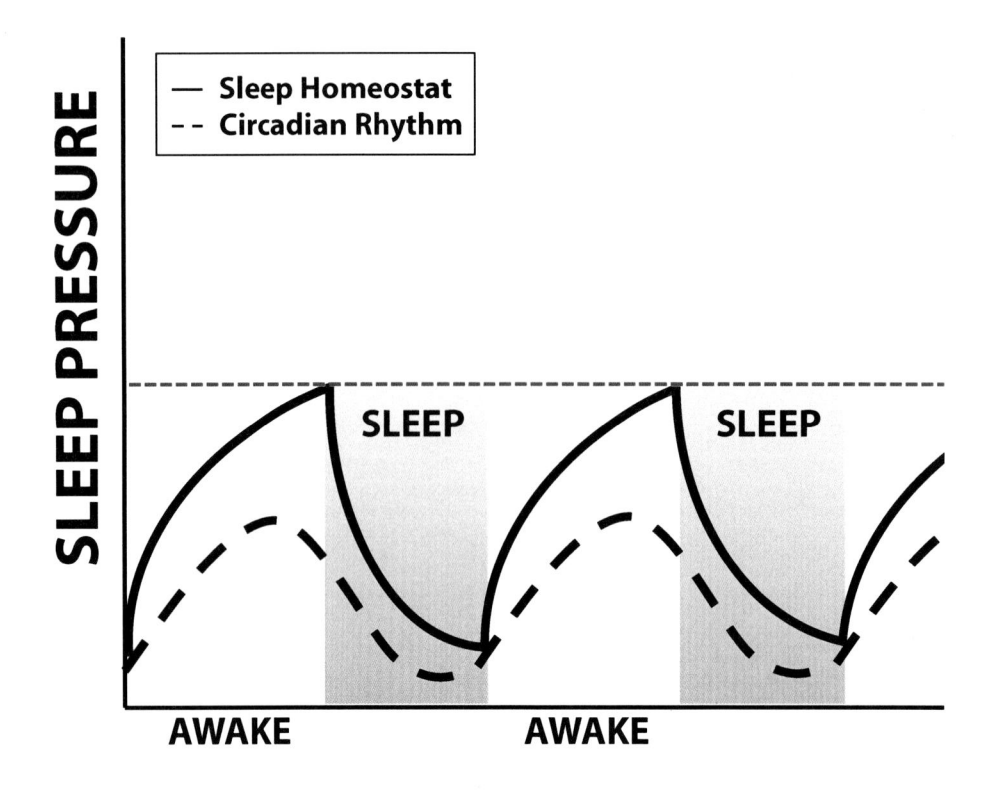

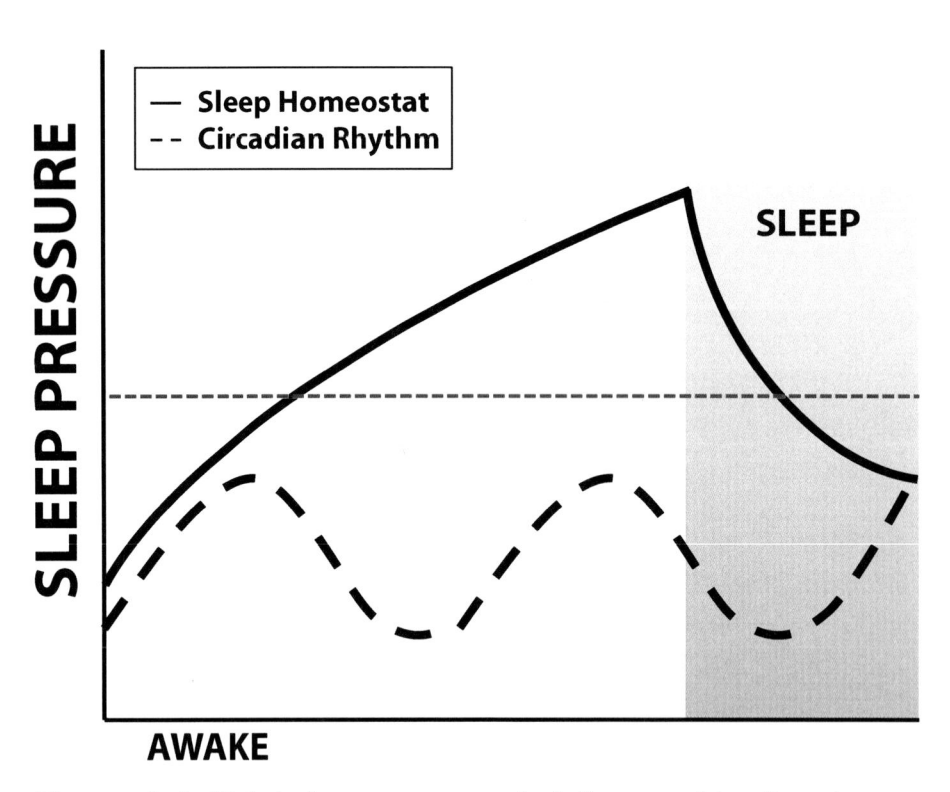

Figure 6-4: Total sleep pressure is influenced by the sleep homeostat and circadian rhythm during normal sleep cycles (top) and extended wake (bottom)

The second factor that regulates sleep is called the **sleep homeostat**. The sleep homeostat keeps track of how long someone has been awake. It also causes sleepiness (also known as sleep pressure) to build up while you are awake and go away when you sleep. Just like a thermostat, which turns on the heat when a room gets too cold, the sleep homeostat helps turn on sleep when a person has been awake too long. The sleep homeostat works by slowly releasing a chemical called adenosine into the brain during wakefulness. The longer a person stays awake, the more adenosine will accumulate in the brain, and the sleepier he will become. Caffeine blocks adenosine signaling, which is why it makes people feel more awake.

> **Sleep Homeostat**: system that keeps track of how long someone has been awake

How do these two systems work together? Adding up sleep pressure from the circadian rhythm and the sleep homeostat determines how sleepy we feel. In the morning, the circadian clock detects that it is time to be awake, and the sleep homeostat has not built up much sleep pressure, so a person will not feel very sleepy. In the evening, the circadian clock communicates that it is time to sleep, and the sleep homeostat has built up a lot of sleep pressure during the day, so together these signals cause very high sleepiness. If you then realize you have a test the next day and stay up all night to study instead of going to bed, you will feel extremely sleepy during the night since sleep pressure from both the homeostat and circadian rhythm is high. As morning approaches, you may actually feel less sleepy because circadian clock is telling you it's time to be awake, even though the homeostat is telling you it's time to sleep. However, extreme sleepiness will return later in the day as the circadian and homeostatic sleep pressure both increase. When you do finally go to sleep, it will be deeper and for longer than normal.

IV. Why do we Sleep?

One of science's greatest mysteries is why we need to sleep. At the moment scientists have several theories, and the evidence indicates that sleep probably has many different functions.

Brain development: As people age, the amount of sleep they get decreases (Figure 6-5). Infants may sleep for 16 hours a day, while adults need far less sleep. A large percentage of infants' sleep is made up of REM sleep, and this percentage decreases with age. This suggests that sleep (especially REM sleep) may help the brain to develop and mature during aging.

Microsleep: a period of sleep lasting only a few seconds that can occur during extreme drowsiness

Attention/focus: Lack of sleep can impair your ability to function during the day. Sleep loss actually has some of the same effects as drinking alcohol! Slowed reaction time, impaired judgment, impaired vision, increased moodiness and aggression, decreased alertness and attention, and difficulty processing and remembering information are all symptoms of sleep loss. In addition, sleep deprivation can cause a person to fall asleep for a few seconds without even realizing it. These brief sleep episodes are called **microsleeps**. It can be very dangerous to perform certain activities, such as driving, while sleep deprived. Someone driving drowsy cannot react as quickly to obstacles or turns in the road. Moreover, if the driver has a 3-second long microsleep while driving at 65 miles per hour, she will travel the length of a football field while asleep! According to the National Highway Traffic Safety Administration, each year in the U.S. at least 1,550 deaths and 71,000 injuries are caused by drowsy driving car crashes. These could be prevented if drivers made sure to get enough sleep.

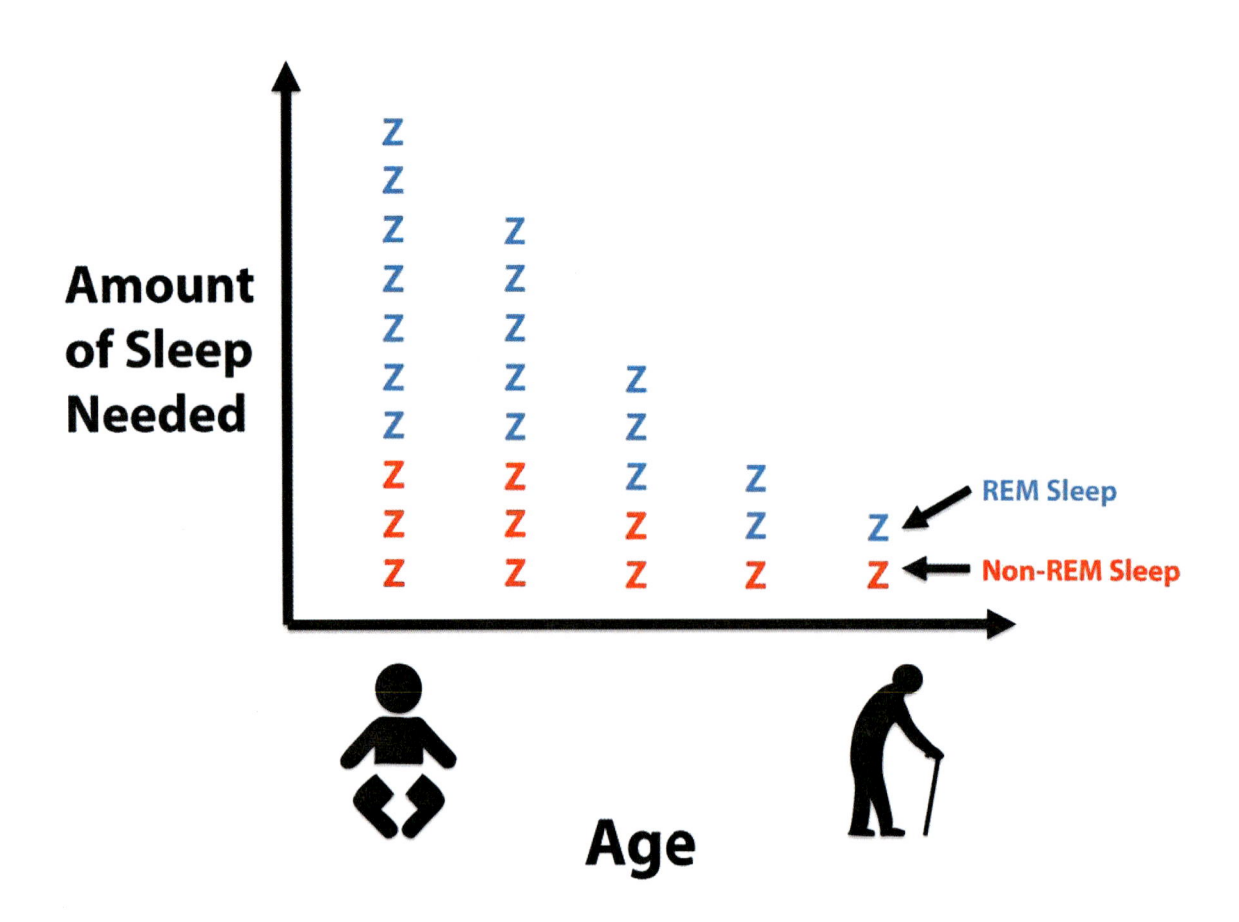

Figure 6-5: *The average duration of sleep, particularly REM sleep, is longest in infants and becomes shorter over the course of a lifetime*

Health: Sleep helps protect you from getting sick. When you do get sick, sleep can help you recover faster. When someone is well-rested, the immune system works much better than when he is sleep deprived. This means that it is harder for the body to fight off infections when it isn't getting enough sleep.

Learning and Memory: Scientists believe that sleep is also important for learning and remembering things. Sleeping after learning something will help you to remember it better the next day. For example, if you study for a test and then get a good night's sleep, you will remember what you studied better than if you stay up all night studying.

V. Sleep Disorders

Sleep is important for many reasons, and not getting enough of it can be very disruptive to our lives. There are many ways in which the systems that control sleep can malfunction, resulting in sleep disorders that seriously impact quality of life. We will discuss a few of these disorders below.

Insomnia is trouble falling asleep or staying asleep. Insomnia can be caused by many things, such as medical, environmental, or psychiatric conditions. For example, being stressed out over an exam, hearing noises outside the bedroom window, or having other sleep disorders can all cause insomnia. Many cases of insomnia, referred to as primary insomnia, have no obvious cause.

> **Insomnia**: trouble falling asleep or staying asleep

Narcolepsy is a disorder in which a person makes uncontrollable transitions from waking to REM sleep. This occurs when the "switch" that controls sleep and wake (shown in Figure 3) does not work properly. The switch malfunctions because the brain doesn't make enough of a neurotransmitter called orexin in the hypothalamus. Normally, when people fall asleep they go into NREM for about 90 minutes before entering REM. People with narcolepsy enter REM almost immediately when they fall asleep.

> **Narcolepsy**: disorder in which the switch in the brain that turns sleep on and off does not work properly

Certain features of REM sleep also occur during waking in narcoleptics. For instance, while they are awake they may experience the atonia normally characteristic of REM sleep. These attacks of paralysis, known as **cataplexy**, can be triggered by emotional experiences such as laughter, surprise, and anger. In addition, while

> **Cataplexy**: attacks of paralysis (the inability to move) in people with narcolepsy, usually caused by emotional experiences

> **Sleep Paralysis**: continuation of REM muscle atonia into waking

> **Hypnagogic Hallucination**: dream occurring while partially awake

> **Obstructive Sleep Apnea**: closure of the airway during sleep, causing short interruptions in breathing

people with narcolepsy are in the process of waking up, atonia may not turn off right away, resulting in **sleep paralysis**. A person experiencing sleep paralysis cannot move for several minutes after waking. Another feature of narcolepsy is the occurrence of **hypnagogic hallucinations**, or dreaming while partially awake.

Obstructive sleep apnea is a disorder in which muscles in the throat periodically obstruct the airway during sleep, causing breathing to stop for at least 10 seconds. These pauses in breathing occur many times throughout the night. This causes people to wake up repeatedly and prevents their bodies from receiving enough oxygen, which can lead to other serious conditions such as heart disease. Obstructive sleep apnea also causes very loud snoring.

Study Questions

1. Describe the differences between NREM and REM sleep.
2. It is very important that atonia can occur during REM sleep but does not occur during other states. Name 2 ways in which atonia can malfunction. In what sleep disorder does atonia malfunction?
3. Staying awake all night will cause you to become very sleepy. However, as morning arrives, you might begin to feel more alert even if you haven't gotten any sleep. Why is this?

Challenge Question

Among people, there is a lot of variability in terms of preferred sleep time per night and preferred times to go to sleep or wake up. This is exemplified by the fact that some of us are "morning people" and some of us would prefer to wake up at noon. Based on the factors controlling sleep discussed in this chapter, what do you think might explain these variations in sleep behavior?

Brain Byte

Have you ever wondered how dolphins keep swimming while they're asleep? In the 1970's, a group of Russian scientists discovered that dolphins and other water-dwelling mammals can sleep with only one hemisphere at a time! This means that one side of a dolphin's brain sleeps while the other half of its brain is still wide awake and able to maintain vigilance. This provided one of many pieces of evidence that in some mammals, brain hemispheres can function highly independently of one another – almost like having two separate brains!

STRESS

Chapter 7: Stress

I. What is Stress?

> **Stress**: a body's physical, mental, and/or emotional reaction to a change in environment, or stressor

> **Stressor**: an external stimulus that demands a physical, mental, and/or emotional reaction from the body

> **Homeostasis**: the ability of the body to maintain a stable internal environment in the face of external or environmental changes

Stress is defined as mental or physical strain or tension. In a neuroscience context, stress is the nervous system's complex response to environmental conditions called **stressors**. Chances are you've experienced stress at several points throughout your life. When we talk about being "stressed out", we're usually referring to the anxious emotions that accompany a change in our environment. In this chapter, we will more closely examine how the brain and the body communicate with one another during stress. We will also discuss the opposite of stress: **homeostasis**. Homeostasis is a state in which the body's internal environment is at its most stable and functional. Finally, we will consider the positive and negative effects of stress on the brain and the rest of the body.

II. The Peripheral Nervous System: Communication between Brain and Body

> **Voluntary Nervous System**: sends messages from the brain to the muscles so that the body can react to sensory information. This is a conscious process.

> **Autonomic Nervous System**: allows the brain to control involuntary bodily functions; it is comprised of the sympathetic and parasympathetic nervous systems.

The peripheral nervous system (PNS) is made up of nerves that transport messages between the body and the central nervous system (recall that the CNS consists of the brain and spinal cord). It is via the PNS that the brain can control the body. The PNS is divided into two branches: the **voluntary nervous system** and the **autonomic nervous system**.

IIA. The Voluntary Nervous System

The voluntary nervous system conveys messages from the brain to the muscles, directing their movement. It is using this system that someone can move purposefully. In the context of the "fight or flight" response, the voluntary nervous system is what directs the muscles of the arm to throw a punch or the muscles of the legs to run away from the scene.

IIB. The Autonomic Nervous System

The autonomic nervous system controls bodily functions that require regulation at the unconscious level, such as breathing, digestion, and heart rate. It is further split into two branches: the **sympathetic** and **parasympathetic nervous systems**. Generally, the sympathetic nervous system promotes "fight or flight," and the parasympathetic nervous system promotes "rest and digest." Table 7-2 lists how the sympathetic and parasympathetic nervous systems affect various organs and body systems.

> **Sympathetic Nervous System**: unconsciously primes the body for action in response to a stressor, as in the "fight or flight" response. This system diverts blood from resting activities such as digestion.

> **Parasympathetic Nervous System**: controls unconscious functions while the body is at rest, as in the "rest and digest" state

	Parasympathetic Division "Rest and Digest"	**Sympathetic Division "Fight or Flight"**
Eyes	Constricts pupil	Dilates pupil
Salivary Glands	Stimulates salivation	Inhibits salivation
Heart	Slows heartbeat	Accelerates heartbeat
Muscles	Reduces blood flow to muscles	Increases blood flow to muscles
Lung	Constricts airway	Relaxes airway
Stomach	Stimulates digestion	Inhibits digestion
Intestines	Stimulates intestinal motility	Inhibits intestinal motility

Table 7-1: Parasympathetic vs Sympathetic Nervous System

III. Acute vs. Chronic Stress

> **Acute Stress**: the body's adaptive response to an immediate perceived threat

> **Chronic Stress**: prolonged activation of the stress response that can have many adverse effects, such as weight gain, inflammation, and disease

> **Neuroendocrine System**: helps the body maintain homeostasis by releasing chemical messengers called hormones that influence bodily functions.

> **Epinephrine**: a hormone that promotes the "fight or flight" response by increasing heart rate and directing energy to the muscles

> **Cortisol**: a hormone that supports replenishment of the body's energy stores and helps the cardiovascular system function efficiently

> **Corticotropin-releasing hormone (CRH)**: a hormone that is released by the hypothalamus in response to stress

The body's response to stress can be divided into two categories based on how long the stressor persists. The first category is called **acute stress**: a stressful event that lasts only a short time. Thousands of years ago, an acute stressor for our ancestors might have been the pressing need to escape from a hungry lion. Today, you might experience acute stress right before you take an important test. This type of stress tends to have beneficial effects on the mind and body.

However, some difficult situations can lead to **chronic stress**: a stressful event or condition that is long-lasting and unrelenting, such as caring for a terminally ill relative. This type of stress can hurt the affected person in the long run, and it may even lead to disease.

IIIA. The Body's Response to Acute Stress

In the modern world, we usually don't have to worry about escaping from lions. However, the body's physiological response to acute stress can still be very helpful. In the face of an immediately "threatening" event, the **neuroendocrine system** releases a combination hormone/neurotransmitter called **epinephrine** (also known as adrenaline) as well as the hormone **cortisol** into the bloodstream.

The neuroendocrine system serves as an interface between the brain and the body: once sensory systems communicate information about a potential threat to the brain, the brain reacts by triggering specialized structures called glands to secrete chemical messengers (like epinephrine and cortisol) that tell the body how to respond to the threat. Epinephrine prepares the body for the "fight or flight" response by increasing heart rate, dilating air passages, and contracting blood vessels.

The branch of the neuroendocrine system that responds to stress is called the hypothalamic-pituitary-adrenal axis, better known as the HPA axis. For a diagram of the HPA axis, see Figure 7-1. In response to threatening events, a brain region called the hypothalamus secretes **corticotrophin-releasing hormone** (CRH), which

HPA axis

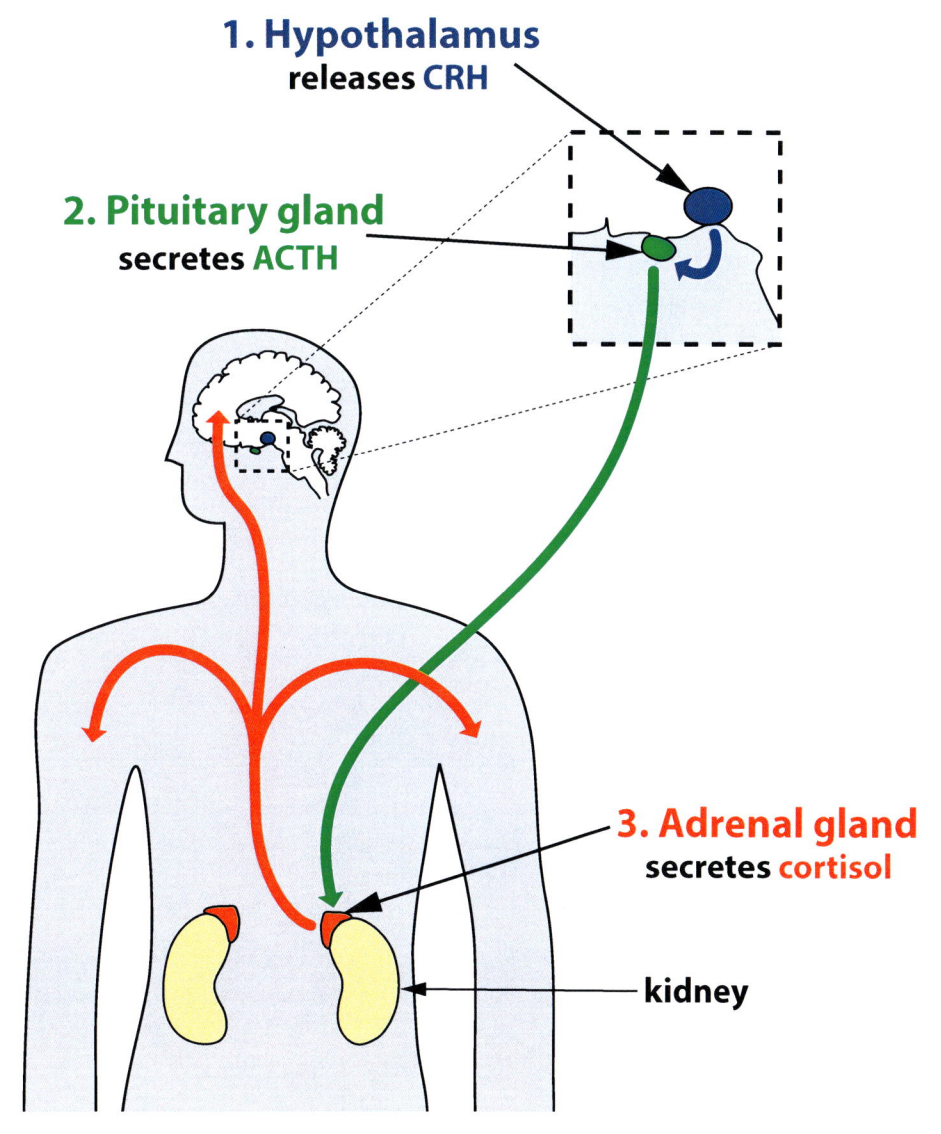

1. Hypothalamus releases CRH

2. Pituitary gland secretes ACTH

3. Adrenal gland secretes cortisol

kidney

Figure 7-1: The Hypothalamic Pituitary Adrenal (HPA) axis

stimulates the pituitary gland to release **adrenocorticotrophin hormone** (ACTH). ACTH in turn triggers the adrenal glands (part of the neuroendocrine system) to release the hormone cortisol.

Cortisol's effects can be both immediate and prolonged. The short-term effects prepare the body to face the situation at hand; these include mobilizing the body's energy stores for quick use, increasing cardiovascular tone, and delaying non-essential functions such as digestion, growth, reproduction, and eating.

Cortisol's slower-acting effects serve to return the body to homeostasis. Homeostasis is important because it keeps your body operating correctly: to be in perfect homeostasis, your body needs

Adrenocorticotropin hormone (ACTH): a hormone that is released by the pituitary gland in response to stress

to maintain a constant temperature and pH, a good balance of nutrients, and the right amount of water in each cell. If your body strays too far from homeostasis for too long, you can get very sick. Therefore, one very important function of cortisol and the stress response in general is to get you back to homeostasis as soon as possible after a stressor.

Cortisol and epinephrine work together to optimize your response to a stressor. First, they help you make clear memories of the stressful situation so that you are sure to remember what happened (and perhaps avoid that situation in the future). Cortisol and epinephrine also increase immune function, and protect the body from pathogens. All of these effects allow the body to deal with a stressor in the most efficient way possible, leaving you ready to dash away from that lion (or at least tackle that tricky algebra problem).

Several brain regions besides the hypothalamus are involved in controlling HPA axis activity (see Figure 7-2). The frontal cortex (an area associated with cognitive processes such as reasoning and planning) sends inhibitory signals to the HPA axis following a

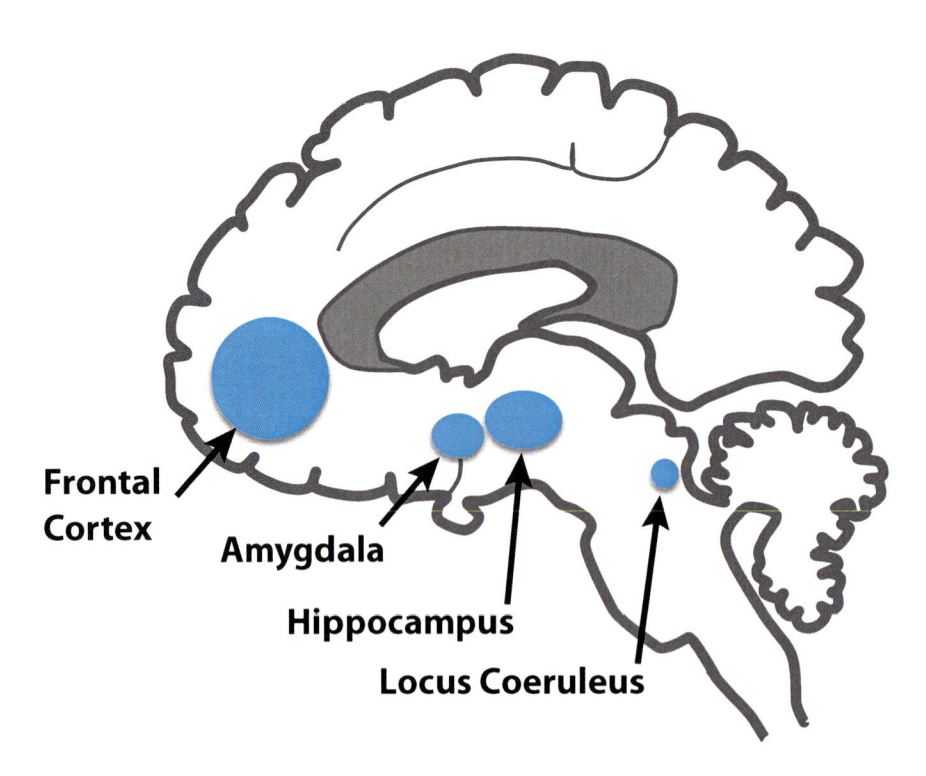

Figure 7-2: Following stress, HPA axis activity is modulated by input from several other brain regions including the frontal cortex, hippocampus, amygdala, and locus coeruleus.

stressor. This helps suppress the stress response once a stressor is removed. The HPA axis also receives negative feedback from the hippocampus, a structure heavily involved in processing of memories.

On the other hand, some brain regions promote activation of the HPA axis. For example, the amygdala, which is involved in emotional processing, is activated during stress. The amygdala sends excitatory signals to the hypothalamus, thereby stimulating the HPA axis. Likewise, the locus coeruleus is involved in initiating the stress response by secreting norepinephrine (similar to epinephrine), which activates the hypothalamus.

IIIB. The Body's Response to Chronic Stress

Prolonged exposure to cortisol can have negative effects on your body and brain. Although it may promote memory-formation in the short-term, chronic cortisol release can actually damage cells in the hippocampus, impairing a person's ability to learn. Long-term exposure to cortisol can also suppress activity of the immune system, increasing the likelihood of illness. Too much cortisol for too long can even impede bone formation, putting someone at increased risk of developing osteoporosis. A number of other diseases are also associated with chronic stress, including heart disease, obesity, anxiety disorders, addiction, and cancer.

IV. The Role of Opioids

In response to stress, neurotransmitters called **opioids** may also be released into the brain and bloodstream from the hypothalamus and pituitary gland, respectively. These chemicals can directly influence cardiovascular (heart) and respiratory (lung) activity as well as centrally decrease anxiety and pain perception.

> **Opioids**: neurotransmitters released into the brain and bloodstream that can counteract many of the acute and chronic effects of stress, promoting relaxation

You might have heard of "runner's high," a feeling of intense pleasure as a result of vigorous exercise. This sensation comes from one of the most well-known classes of opioids, the endorphins. The calming effects of meditation may also come from release of endorphins. For these reasons (and many others), exercise and meditation are often helpful to people who experience chronic stress.

V. Stress/Anxiety Disorders

Although sometimes unpleasant, stress is usually a natural and healthy process that helps the body to maintain homeostasis in a changing environment. Recall that homeostasis is the ability of the body to maintain a stable, internal environment in the face of external or environmental changes. An example of a simple, homeostatic process is when the body sweats during a hot summer day. By sweating, the body is able to cool itself and maintain a constant internal temperature despite increased outside temperatures. Chronic stress, genetic factors, and traumatic experiences can sometimes lead to maladaptive stress patterns or disorders that can seriously disrupt homeostasis. We will discuss a few of these disorders below.

Post-traumatic stress disorder (PTSD) is diagnosed in people who experience hyperarousal and recurring flashbacks or nightmares following a traumatic event. Imbalances of the HPA axis are thought to play a critical role in this disorder, which can persist for months or years.

Obsessive-compulsive disorder (OCD) is a disease in which the patient has uncontrollable obsessions (germs, intrusive thoughts, disorganization, or others) as well as compulsive behaviors (washing, counting, checking to make sure appliances are turned off) that ease the anxiety associated with the obsession. For instance, those who are obsessed with germs may have compulsions such as washing their hands several times in a row, or refusing to touch anything in a public space. Although the compulsions do temporarily reduce anxiety, they are usually irrational responses that can intrude upon daily life.

Generalized anxiety disorder refers to a state of excessive chronic anxiety that is irrational and uncontrollable. This can often lead to sleep disturbances, irritability, and an inability to concentrate on anything other than the source of the anxiety.

In all of these disorders, stress can aggravate symptoms, but symptoms can also increase stress. Without treatment, this vicious cycle can seriously affect quality of life. It's important for everybody (not just those with anxiety disorders) to try to minimize the effects of chronic stress. Some good strategies are to make sure to eat well, get enough sleep, and spend quality time with loved ones. Can you think of other ways to reduce the negative effects of stress on your life? Your body and brain will thank you!

Study Questions

1. What is the difference between the voluntary and autonomic nervous systems?
2. What is the difference between the sympathetic and parasympathetic nervous systems?
3. Name two stress hormones.
4. Name three examples of "stressors". Which of these would most likely cause acute stress? Which might cause chronic stress?

Challenge Question

Though we all experience stress in one way or another, some of us tend to be more affected by stressful situations than others. Based on what you've read in this chapter, what are some factors that may cause someone to be more or less sensitive to the effects of stress?

Brain Byte

Even though it's not fun to be stressed out about an exam, a century of research indicates that experiencing a low level of stress actually improves performance. This is often referred to as the Yerkes-Dodson Law, which states that as stress levels increase, cognitive abilities do as well, until a certain point. If stress levels become too high, cognitive abilities then begin to decline.

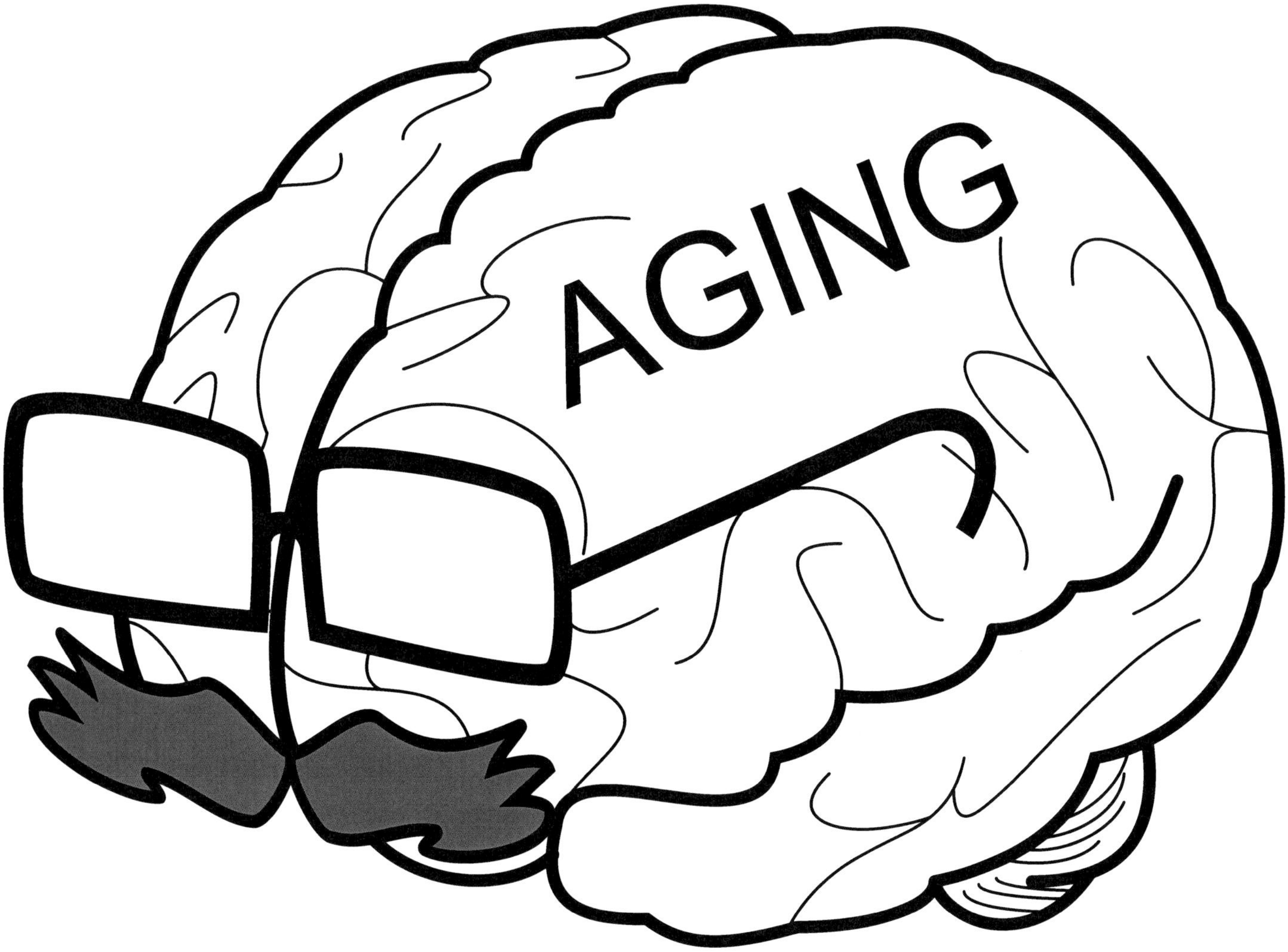
AGING

Chapter 8: Aging

I. Aging in the Brain

Think back to the last time you got a paper cut. How long did it take for the cut to heal? For most people, skin on the injured area heals within a few hours. This is due to the renewal or rebirth of skin cells. These cells regenerate your entire top layer of skin every two or three weeks.

Unlike skin cells and many other types of cells in your body, most neurons do not regenerate. In fact, many of the neurons that you have at birth will continue to grow and develop with you throughout your life. In other words, neurons are among the oldest cells in your body and about the same age as yourself. Because **neurogenesis**—the birth of new neurons—slows down dramatically after birth and declines further with age, preserving the health of the existing neurons in your brain is amazingly important.

> **Neurogenesis**: the process of creating new neurons, which is most active during pre-natal development (before birth)

To understand how the brain ages, you first need to understand what **aging** is. In every day terms, aging means the growth of an organism over time. For your brain, this means the accumulation of small changes in neurons and molecules over the course of your life.

> **Aging**: an accumulation of changes in a cell or an organism that develop over time

IA. What Causes the Brain to Age?

One example of a cellular change that contributes to neuronal aging is the buildup of **reactive oxygen species** (ROS). ROS are produced by normal metabolism of oxygen and are required to produce energy for your cells. They also serve as important signaling molecules in neurons. However, environmental stress or cellular growth over time can lead to an increase in ROS that is harmful to neurons. Build up of ROS can actually damage DNA, RNA, and proteins in the cell, all of which contribute to cellular aging.

> **Reactive Oxygen Species (ROS)**: chemically reactive molecules containing oxygen. ROS are produced by normal metabolism of oxygen and can act as cellular signaling molecules.

IB. Age-related Changes in the Brain

How does aging affect the brain? There are both **cognitive** and **structural** changes in the brain associated with aging. Cognitive abilities—mental processes related to thinking—may improve or decline with aging. Think about elderly, healthy adults that you know. They may be excellent story-tellers or good at giving advice. Not surprisingly, the many life experiences that come with old age increase knowledge and understanding. However, you might alsot notice that older adults have problems with memory, such as remembering names of people they have met. The following table (8-1) summarizes normal aging related changes in cognition:

Cognitive Function: mental abilities that require thought, such as applying knowledge to solve a problem or using language

Structural Organization: divisions of the brain or groups of cells that ordinarily work together to serve a particular function.

Cognitive Abilities that DECLINE with Age	Cognitive Abilities that are PRESERVED or IMPROVE with Age
Ability to solve new problems or learn new skills	Ability to remember accumulated knowledge and experiences
Visual-spatial abilities such as navigating (like following driving directions)	Ability to use language (although vocabulary recall may decrease)
Ability to recall memories, form new memories or keep information in short-term memory	Executive function, ability to plan and make decisions

Table 8-1: Changes in cognitive abilities with aging

Many of these cognitive changes, such as forgetfulness and difficulties with spatial navigation, can be explained biologically by the structural changes in the brain associated with aging. With healthy aging, the brain may lose some neurons over a lifetime, but there is no widespread or massive loss of neurons. It is important to note the difference between normal aging and neurodegeneration. We will discuss neurodegeneration later in this chapter. During the process of healthy aging, normal structural changes that occur gradually include the following:

1) The hippocampus gets smaller. Why is the hippocampus so essential? As you may recall from Chapter 5, the hippocampus is very important for learning and memory. Shrinkage of the hippocampus

can result in memory loss and difficulties in spatial memory or spatial navigation as you age.

2) Due to the loss of some neurons, your brain tissue no longer takes up as much space. However, the size of your skull remains the same. As a result, the spaces or **ventricles** in and around the smaller brain areas get larger. Ventricles are filled with cerebrospinal fluid. This fluid cushions your brain and protects it from injury.

3) The appearance and size of the brain's **white matter** changes. Recall that white matter is made up of axons that connect brain regions to one another. Decreases in white matter cause the connections between brain regions involved in memory, language, and visual-spatial tasks (like following driving directions) to slow down.

To see what some of these changes look like, see Figure 8-1.

Ventricles: fluid filled spaces within the brain that circulate cerebrospinal fluid and nutrients to tissue

White Matter: myelinated axons that connect neurons to one another

Young Brain | **Normal Aging Brain**

Larger Ventricles

Smaller Hippocampus

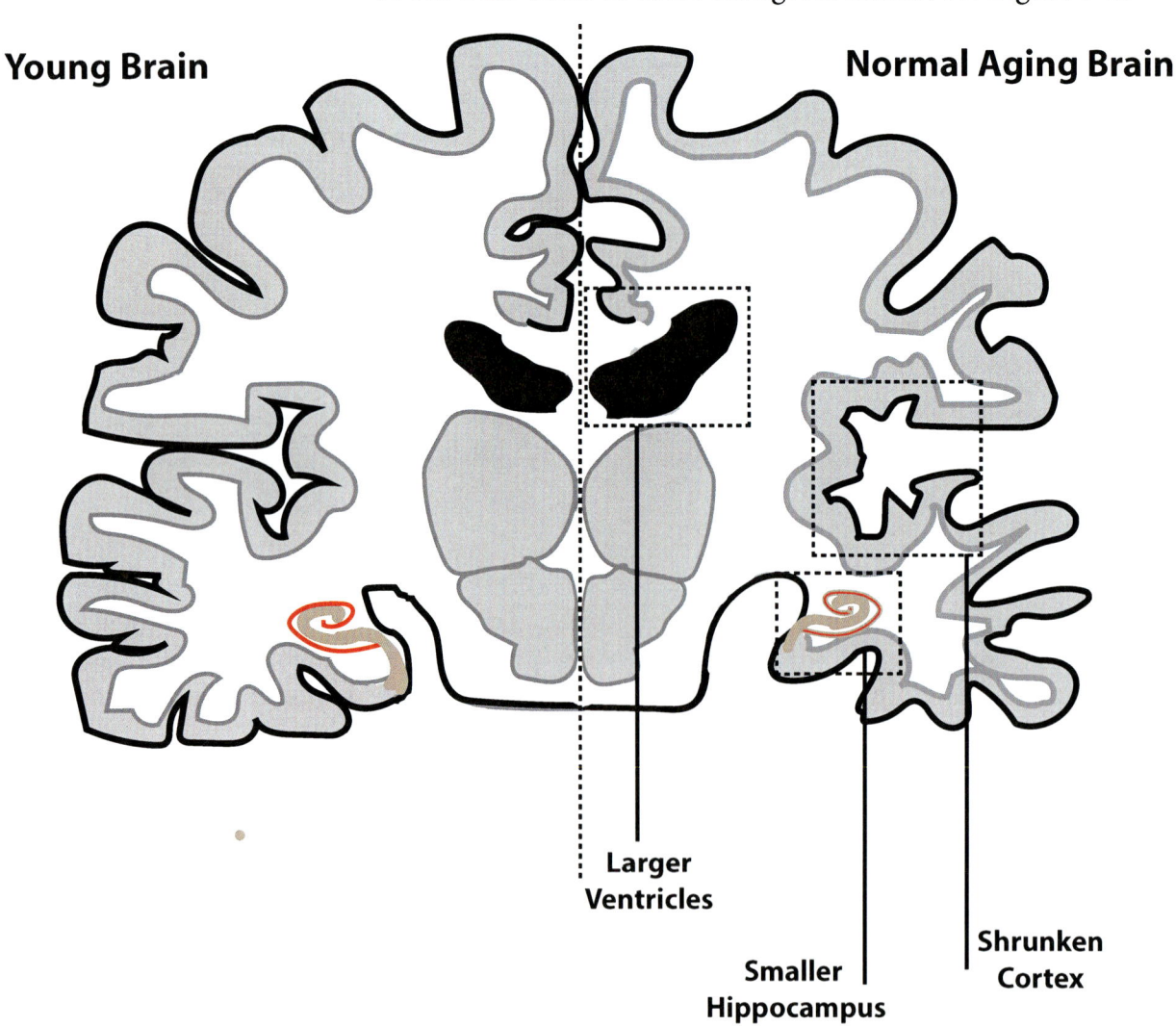

Shrunken Cortex

Figure 8-1: Comparison of a normal young brain (left) with a normal aging brain (right)

IC. Additional Factors that Influence Aging

What influences how quickly the brain ages? Similar to the aging of other organs, both genetics and environmental factors play important roles. In part, the genes that you inherited from your parents will factor into your aging. Twin studies have shown us that identical twins raised together are more likely to have similar life spans and age related changes in cognition compared to other siblings in the family. Certain genes are beneficial to increasing your lifespan and the health of your brain, while others may be harmful, increasing the risk for **stroke** or **Alzheimer's disease**, for example. There can also be genetic mutations that are not inherited, but instead are caused by the environment, such as exposure to toxins.

Other environmental factors that do not change your genes can still influence how your brain ages. For example, low achievement in education can quicken memory loss in old age and can increase chances for developing dementia (which is a good reason to do your homework!). Choices in diet and smoking can increase the likelihood of developing diseases like diabetes, hypertension (elevated blood pressure), and arteriosclerosis (thickening of the walls of a blood vessel due to fat buildup), which all have damaging effects in the brain. In summary, both nature (your genes) and nurture (your environment and behavioral choices) influence how quickly your brain will age.

> **Stroke**: very fast loss of brain functions as a result of loss of blood to the brain

> **Alzheimer's disease**: a neurodegenerative disease that primarily affects the hippocampus and cortex. Plaques and tangles are the cellular markers of Alzheimer's disease.

II. Healthy Brain Aging

Increases in the average life expectancy highlight the importance of healthy aging. Consider this: 100 years ago, the average life expectancy in the United States was less than 50 years. Today, Americans have an average life expectancy of 78 years. That means we are living a quarter of a century longer than our ancestors did one century ago! This is a great reason to do everything you can to make sure your brain stays healthy into old age. Research on brain aging has offered us several strategies to delay the effects of aging and prevent cognitive decline.

Additive Strategies: tips for healthy brain aging to strengthen connections between neurons in your brain and increase structural plasticity to keep your brain functioning smoothly.

- Stay intellectually engaged by reading books and newspapers or by working on activities that require strategy like crossword puzzles or strategic card games.

- Surround yourself with family and friends to stay socially active.

- Exercise regularly.

Preventative Strategies: tips for healthy brain aging to help protect you against neuronal injury.

- Try to decrease your stress level.

- Eat a healthy diet. Antioxidants in berries, leafy green vegetables, and nuts help protect your neurons from damage.

Our understanding of how the human brain ages is constantly changing as medical advances allow people to live longer and longer. By the time that you are in your 60s or 70s, there may be a whole new set of tools to help your brain age gracefully.

III. Disease in the Aging Brain

Risk Factor: a variable that is associated with an increased risk for a specific disease, but does not necessarily cause the disease itself. For example, doctors cannot say that smoking causes strokes, but smokers are at higher risk for having a stroke.

Blood Clot: also known as a thrombus, this is a buildup of platelets in the blood that can block normal blood from flowing through a blood vessel

Why do you want to protect your brain as it ages? Old age is the top **risk factor** for age-related neurological disorders such as stroke and Alzheimer's disease. Both stroke and Alzheimer's disease cause the death of neurons. We call these types of diseases "neurodegenerative diseases". Unlike the normal brain aging described above, neurodegenerative diseases can cause massive, widespread loss of neurons.

IIIA. Stroke

A **stroke** occurs when blood flow to the brain is disrupted. This could be caused by a **blood clot** or by bleeding in the brain (see Figure 8-2). The brain cannot function without oxygen carried by blood. Depending on where the stroke is located, different neurons will be affected and the patient will experience different symptoms. For example, if the stroke occurs in the patient's motor cortex, he may be unable to move one or more limbs on one side of his body.

Under certain circumstances, patients can receive treatment for a stroke such as a drug to dissolve the blood clot or surgery to repair a damaged blood vessel. If the treatment is performed quickly and successfully, the patient may have no permanent damage. However, stroke can rapidly cause neuronal death and lead to permanent neurological disability. Early recognition of the symptoms of stroke and fast treatment can help prevent long-lasting injury and neurodegeneration.

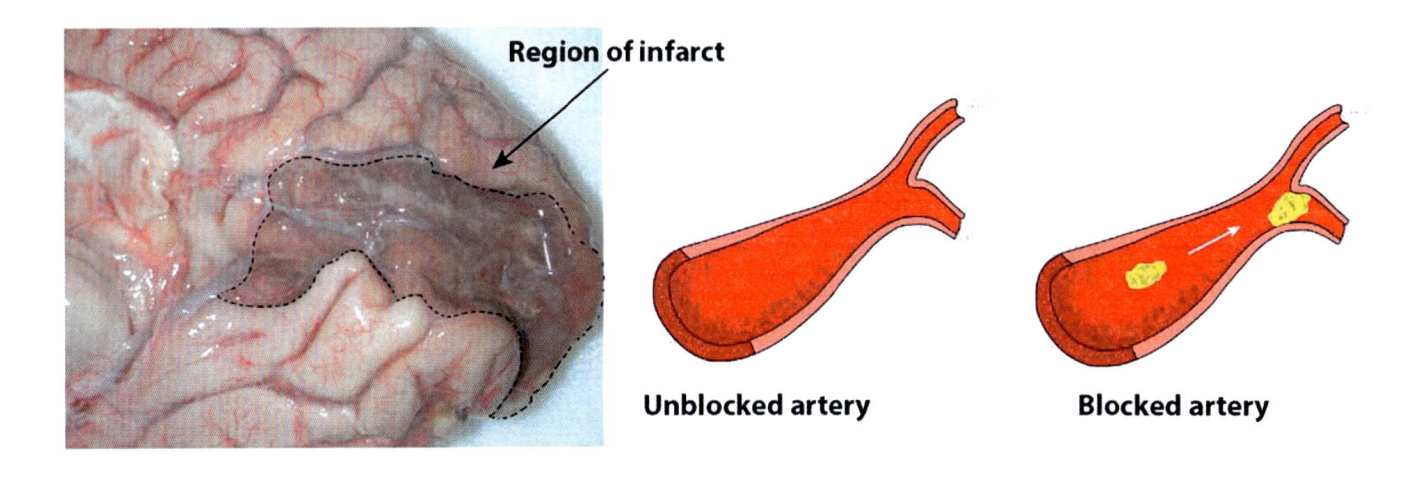

Figure 8-2: Photograph of a brain with a massive stroke in the occipital lobe. Stroke can be caused by a blockage in a blood vessel (far right).

IIIB. Alzheimer's disease

Alzheimer's disease is the most common form of **dementia**. It is most often diagnosed in patients over 65 years old, although certain genetic risk factors can lead to onset of the disease earlier in life. Patients with Alzheimer's disease suffer from episodic memory loss (recall from Chapter 5 that this is memory for events), confusion, changes in mood, and, late in the disease, breakdown in language skills. While the cause of Alzheimer's disease is not known, research has shown that the disease is linked with buildup of proteins, both inside of neurons (called tangles) and in the areas in between cells (called plaques). These tangles and plaques are linked to loss of neurons and synapses, especially in the hippocampus and frontal cortex. Both these brain areas become dramatically smaller over the course of the disease. To see what a brain scan of a patient with Alzheimer's disease looks like, see Figure 8-3.

> **Dementia**: loss of certain cognitive functions such as memory, thinking, language and behavior. It is associated with certain neurological disorders.

Traumatic Brain Injury: caused by an external force injuring the brain, such as a car accident or severe fall. Concussions are the most common type of traumatic brain injury.

Amyotrophic Lateral Sclerosis (ALS): a neurodegenerative disease that leads to the death of motor neurons

Each neurodegenerative disease has different causes and distinct pathology. However, research into why neurons die in one disease might help researchers who are studying a completely different disease. For example, some professional athletes in high-contact sports (like football or boxing) suffer from repeated traumatic brain injury. Scientists investigating this kind of brain damage are now comparing **traumatic brain injury** with the motor neuron disease **amyotrophic lateral sclerosis** (ALS) to better understand both conditions and develop treatments.

Although there are very few if any cures available today for neurodegenerative diseases, future research into these disorders will hopefully lead to the development of new therapies and preventative strategies to keep our brains healthy into old age.

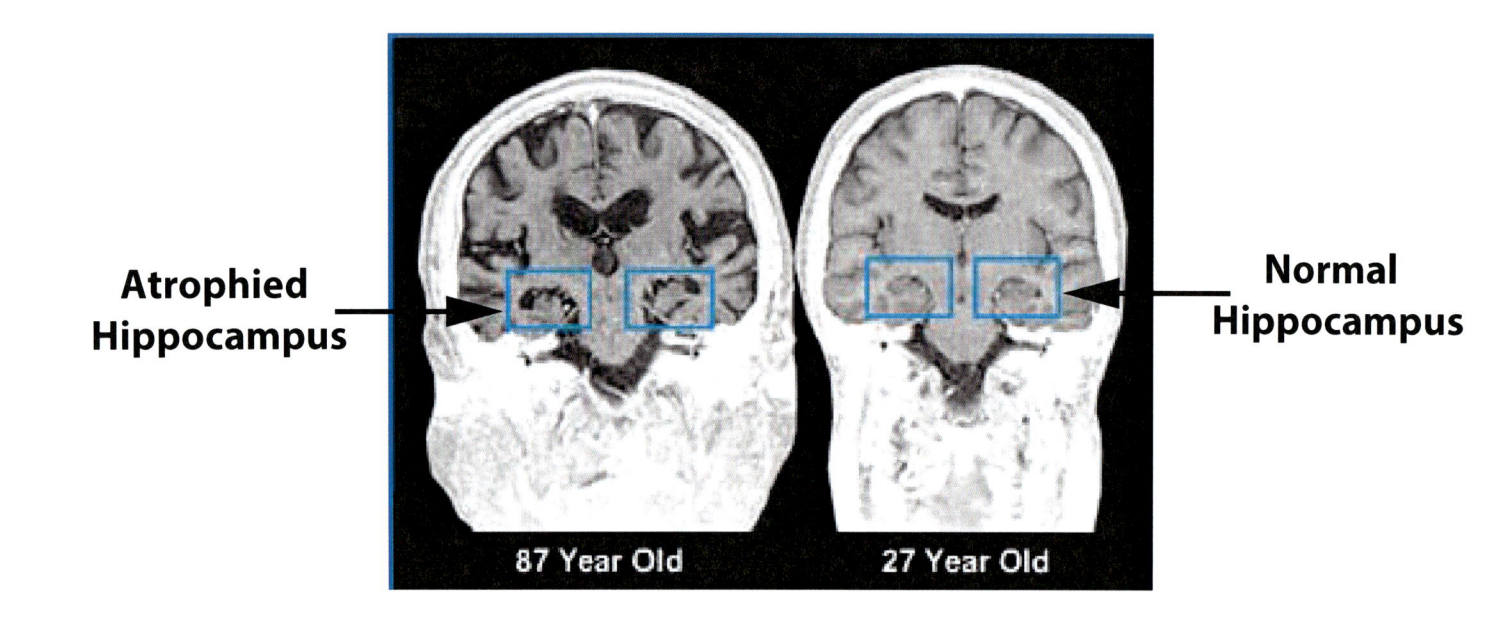

Figure 8-3: Comparison of the size of the hippocampus in a Magnetic Resonance Image of a young brain (right) and an Alzheimer's brain (left)

Study Questions

1. List 1 type of cognitive change and 1 type of structural change that occurs during normal aging of the brain.
2. What is the top risk factor for neurological disorders such as stroke and Alzheimer's disease?
3. Remember that different brain structures have specialized functions. If a stroke occurs in a particular brain region, it will interfere with the normal function of that area. Below is a list of brain areas and cognitive impairments that might occur with stroke. Match the injured brain area with the cognitive impairment that you might expect. In order to answer this question, think about what you learned in previous chapters.

Brain area with damage
A. Hippocampus
B. Left motor cortex
C. Occipital lobe

Cognitive impairment
Visual hallucinations   ____
Inability to form new memories   ____
Paralysis of the right arm   ____

Challenge Question

Do you think it is possible for patients with stroke to have episodic memory loss? If so, will these patients display a similar clinical phenotype to patients with Alzheimer's disease? Why or why not?

Brain Byte

A diagnosis of "mild cognitive impairment" fills the gap between healthy aging (in which most brain function is preserved) and Alzheimer's disease (in which many brain functions are severely impaired). Patients with mild cognitive impairment are not as disabled as patients with Alzheimer's, but they do have an increased risk of eventually developing this devastating disease. Mild cognitive impairment patients are unable to encode and retrieve contextual aspects of memory. This means they can't remember relations between items but can remember individual items. Many scientists are studying Mild Cognitive Impairment with the goal of developing early treatments for dementia.

ADDICTION

Chapter 9: Drug Addiction

I. Drug abuse is a serious health problem.

> **Drug Abuse**: harmful drug misuse or excessive use

Drug abuse is a serious public health problem in the United States and other countries around the world. About 10% of Americans abuse drugs on a regular basis, and abuse of drugs causes around 40 million serious illnesses or injuries every year. The cost of drug abuse, including the costs of health care, crime, and lost productivity is estimated to be over half of a trillion dollars each year in the U.S. One reason that drug abuse is such a serious threat to public health is that it often affects people other than the ones abusing drugs, through drunk or drugged driving, violence, child abuse, and other criminal activity.

The serious problems associated with drug abuse are not limited to the abuse of illegal drugs: both alcohol and tobacco have a strong negative impact on public health as well. Alcohol abuse and addiction kill tens of thousands of people in the US every year, and use by pregnant women can cause birth defects and mental handicaps in their children. Tobacco (smoking) is the leading preventable cause of death in the US, killing nearly half a million people each year.

II. Why do people take drugs?

> **Stimulants**: drugs that increase mental (e.g. wakefulness, alertness) or physical (e.g. blood pressure, heart rate, movement) function

> **Depressants**: drugs that decrease mental or physical function

With such dire negative consequences, it may seem surprising that people abuse drugs in the first place. While the initial decision to experiment with drugs may be made for a variety of reasons, drug use and abuse persist because drugs of abuse are rewarding. They cause pleasure and/or reduce the effects of emotional and physical stress or pain.

Drugs of abuse include both **stimulants** and **depressants**. Stimulants, such as cocaine and amphetamine, are drugs that increase physical and mental functioning. They increase heart rate, blood pressure, wakefulness, and alertness. Depressants, such as heroin and alcohol, decrease physical and mental functioning and cause a feeling of relaxation. Note that depressants do not cause a "depressed" mood. Other drugs, such as ecstasy, LSD, and marijuana, have more complex effects that may include hallucinations. Finally, drugs such as nicotine and ketamine can be both stimulants and depressants depending on the dose.

All addictive drugs induce their pleasant effects by hijacking a system in your brain that was built to respond to **natural rewards**. This system, the brain reward circuit, evolved to encourage productive behaviors such as feeding and procreation by encouraging people (and animals) to repeat these behaviors in order to experience pleasure. Drugs of abuse activate this system inappropriately, causing the drug user to learn that continued taking of the drug produces the rewarding feeling. Over time, drugs of abuse cause long-term changes in the circuitry of the brain. This makes it increasingly difficult for the drug user to stop taking the drug and eventually leads to **drug dependence**. At this point, the user is addicted and requires the drug simply for normal functioning.

III. Drug abuse is a cycle.

Drug addiction is a disorder that goes beyond just drug abuse. True drug addiction involves a pathological desire for the drug that causes compulsive drug-taking and drug-seeking behavior despite negative consequences. People who are addicted to drugs also have trouble controlling the frequency of use and experience extreme difficulty in stopping their drug use despite wanting very badly to quit. Drug addiction often involves a gradual increase in drug dose as the addicted person becomes more **tolerant** to a drug, experiencing less of its pleasurable effects with a dose that was previously very effective. The addicted person must take higher and higher doses of the drug in order to activate the reward pathway.

Drug addiction therefore often exists as a cycle (see Figure 9-1) in which long-term drug abuse leads to alternating attempts to quit (temporary **abstinence**), followed by **relapse**, or the return to drug abuse. Relapse can be caused by many factors, including **withdrawal**, the negative symptoms that occur when drug use is stopped. Relapse can also be triggered by stress, environmental cues associated with the drug, or re-exposure to the drug, even after periods of abstinence lasting several years. Some addicted people do

> **Natural Reward**: substances or activities (like food or sex) that are part of healthy functioning and cause enjoyment. Natural rewards cause dopamine release in the brain.

> **Drug Dependence**: the need to take a drug for "normal" bodily functioning

> **Drug Addiction**: compulsive drug-seeking and drug-taking behavior that persists in spite of negative consequences

> **Tolerance**: the need for higher and higher drug doses to achieve the same effect(s)

> **Abstinence**: no drug use for some period of time

> **Relapse**: the return to drug abuse after a period of abstinence; often due to stress, an environmental cue, or re-exposure to the drug

> **Withdrawal**: the adverse bodily symptoms that occur after stopping use of a drug on which a person is dependent; can prompt relapse to drug abuse

manage to quit, but they must be careful for the rest of their lives to avoid situations that may trigger relapse.

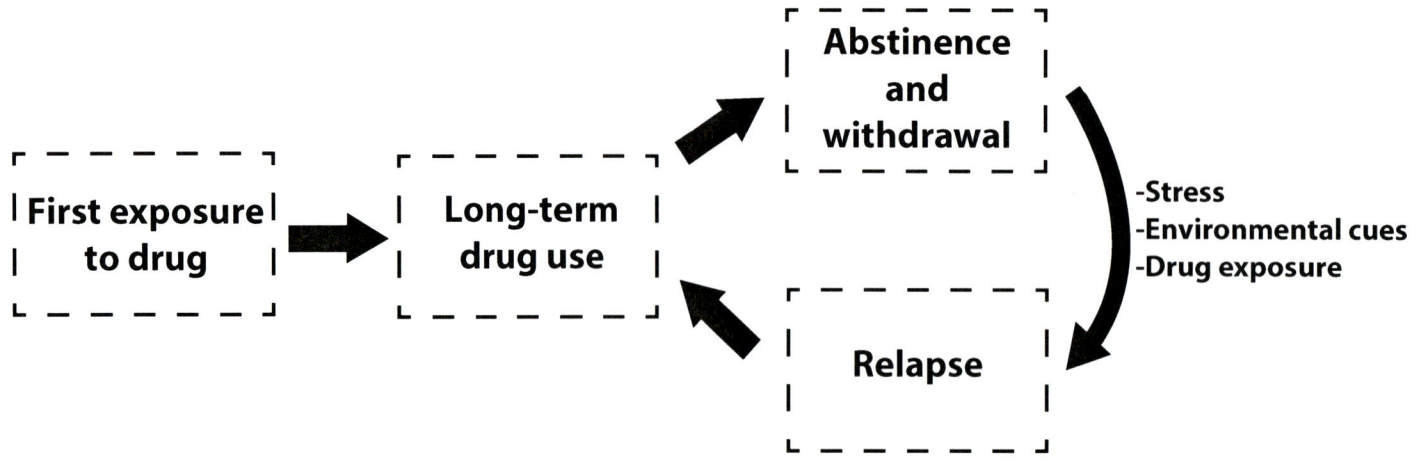

Figure 9-1: *The cycle of addiction*

IV. How Drugs Affect the Brain

As mentioned above, drugs of abuse hijack the brain's natural reward system. This reward system is primarily two brain areas: the ventral tegmental area (VTA) and the nucleus accumbens (NAc). Figure 9-2 will help you visualize these brain areas and help you understand how they are connected to other brain areas affected by addiction. In response to a natural reward, neurons in the VTA release **dopamine** into the NAc. Dopamine is often referred to as the "pleasure neurotransmitter," but what it really does is promote a very specific type of learning: it teaches the brain to repeat an activity that had a positive outcome.

> **Dopamine**: a neurotransmitter that is released in brain reward areas in response to rewarding substances or activities; can be both inhibitory and excitatory

In the case of natural rewards, dopamine release ensures that you repeat behaviors like eating when you are hungry and drinking when you are thirsty. Drugs of abuse, however, cause even more dopamine to be released in the NAc than natural rewards do. This causes an increase in transmission of **glutamate**, the principal excitatory neurotransmitter in the brain. The combined effect is that the user learns very well to repeat the action, which in this case is taking the drug.

> **Glutamate**: major excitatory neurotransmitter of the nervous system; made and released throughout the brain

The VTA also has connections to other areas of the brain, including the hippocampus, amygdala, and prefrontal cortex. All of these areas are components of the limbic system (your emotional brain).

Specifically, the hippocampus is involved in processing environmental or discrete cues related to drugs, and the amygdala is involved in processing the stress-related component of drug-taking behavior.

Over time, drugs of abuse cause long-term changes in the brain. These changes may include altering the firing patterns of VTA neurons, manipulating how dopamine is released in the limbic system, or changing the number and sensitivity of neurotransmitter receptors.

Addiction Circuitry

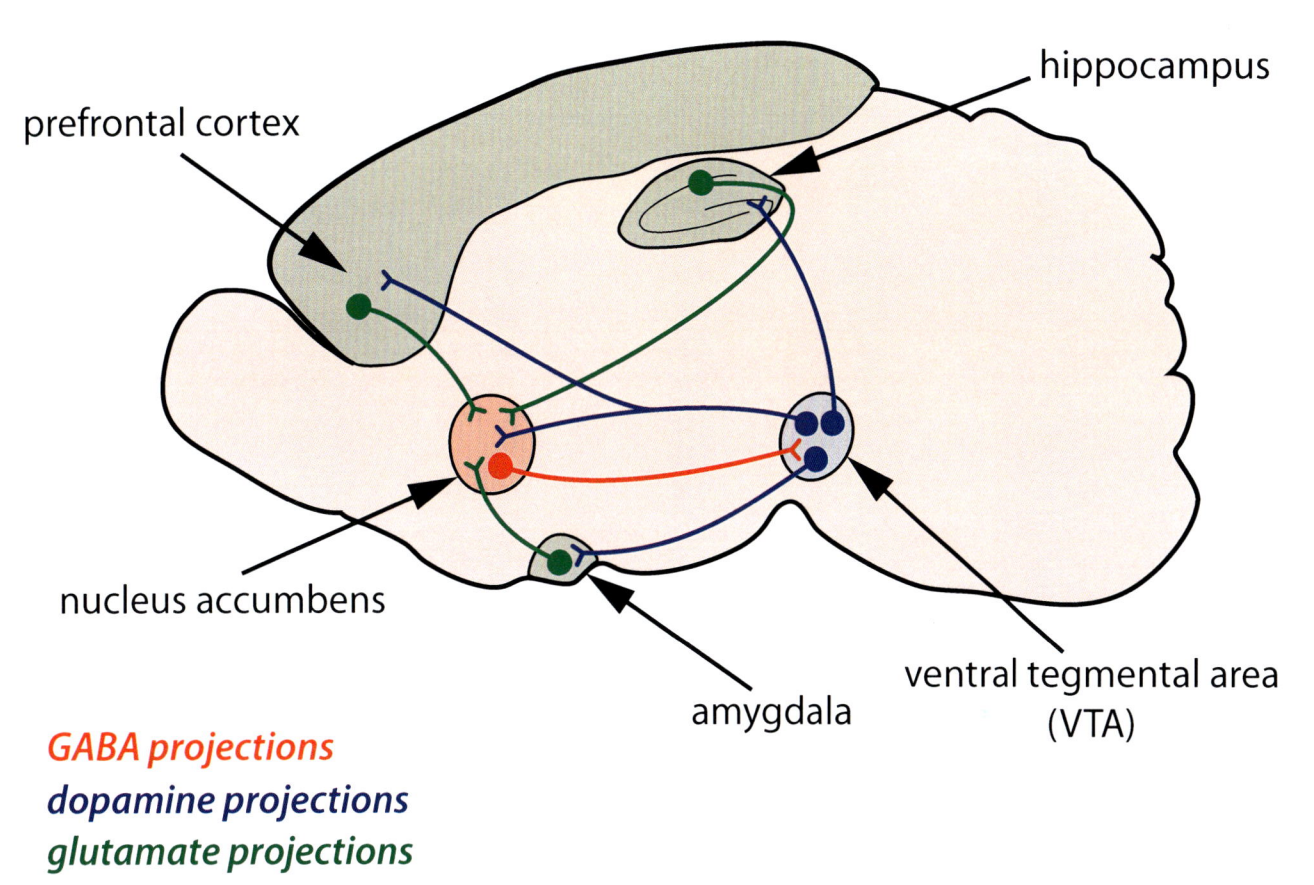

GABA projections
dopamine projections
glutamate projections

Figure 9-2: Addiction circuitry in the limbic system

Even though all drugs of abuse cause dopamine release in the NAc, they do so in many different ways. Below are some examples of how various drugs of abuse cause increased dopamine release in the NAc.

At a normal synapse when no drugs are present, the VTA receives a signal from a natural reward to release dopamine. The dopamine travels across the synapse to bind dopamine receptors in the NAc. The VTA then uses something called a "dopamine transporter" to remove dopamine from the synapse. This ensures that dopamine cannot continue to bind dopamine receptors over and over. This is shown in Figure 9-3.

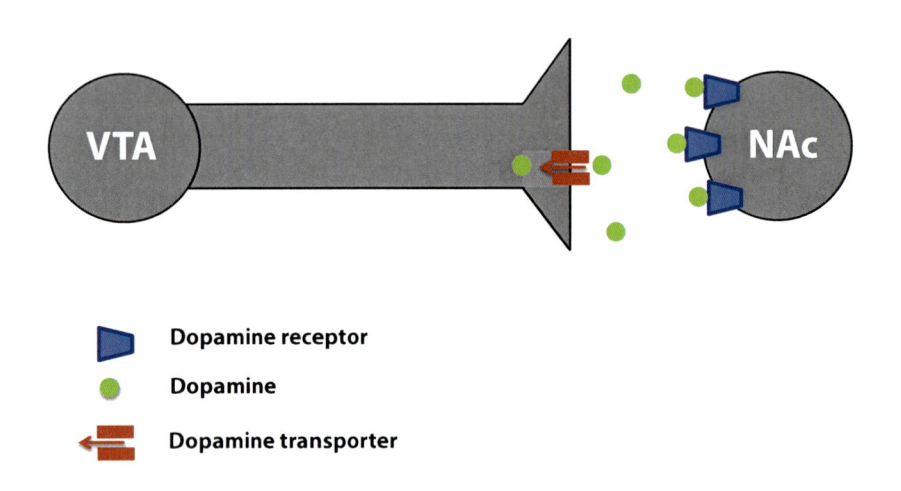

Figure 9-3: Normal synapse: Natural rewards stimulate the VTA to release dopamine

Cocaine blocks the dopamine transporter, which is normally responsible for removing dopamine from the synapse after it is released. This results in more dopamine remaining in the synapse for a long time, binding over and over to receptors in the NAc (Figure 9-4).

Amphetamine also affects the dopamine transporter. Unlike cocaine, it reverses the actions of the transporter, causing it to pump dopamine into the synapse rather than remove it. The result is similar to that of cocaine: dopamine is present in synapses in the NAc for longer than normal (Figure 9-5).

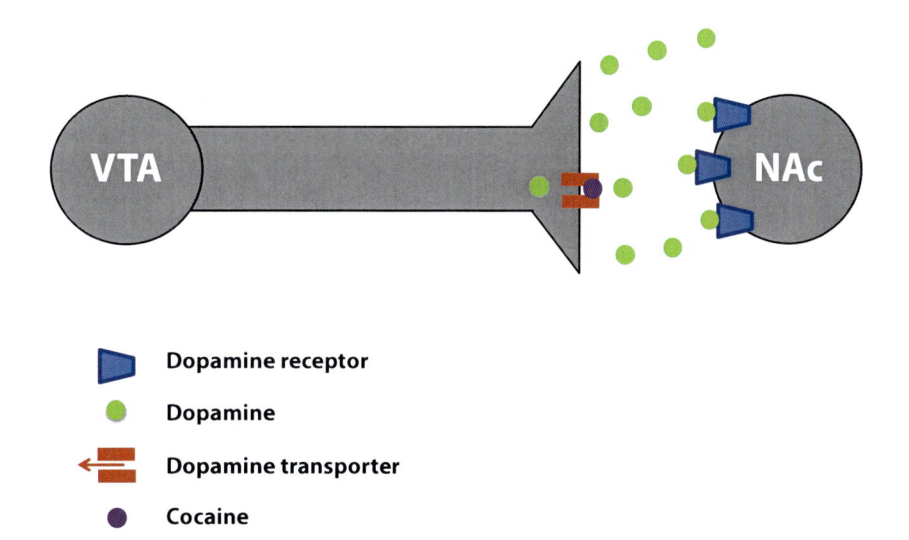

Figure 9-4: *Cocaine blocks the dopamine transporter, which is normally responsible for removing dopamine from synapses after it is released*

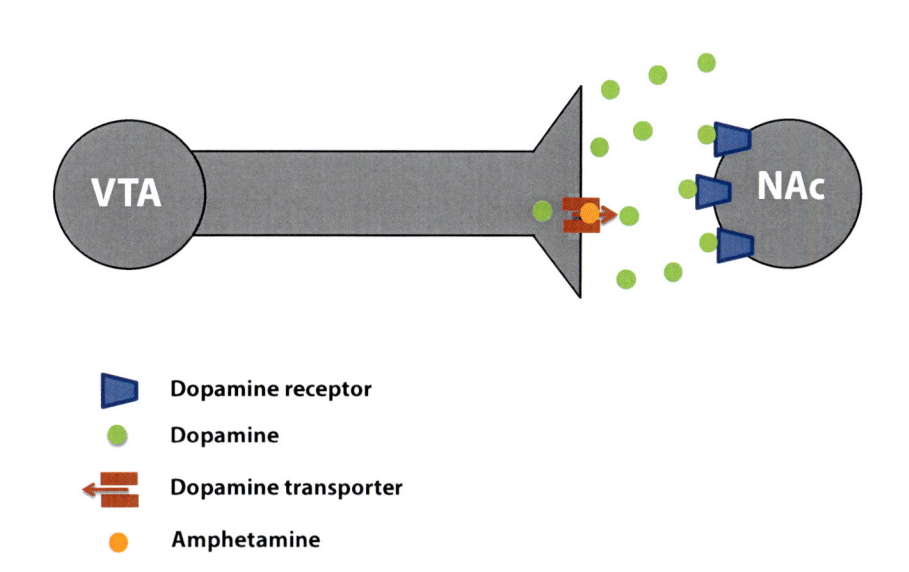

Figure 9-5: *Amphetamine reverses the actions of the dopamine transporter so that it pumps dopamine into the synapse rather than removing it*

Nicotine binds to acetylcholine receptors on the axon terminals of VTA neurons releasing dopamine. This causes a greater depolarization and therefore a greater release of dopamine into the synapse (Figure 9-6).

Dopamine receptor
Dopamine
Dopamine transporter
Acetylcholine receptor
Nicotine

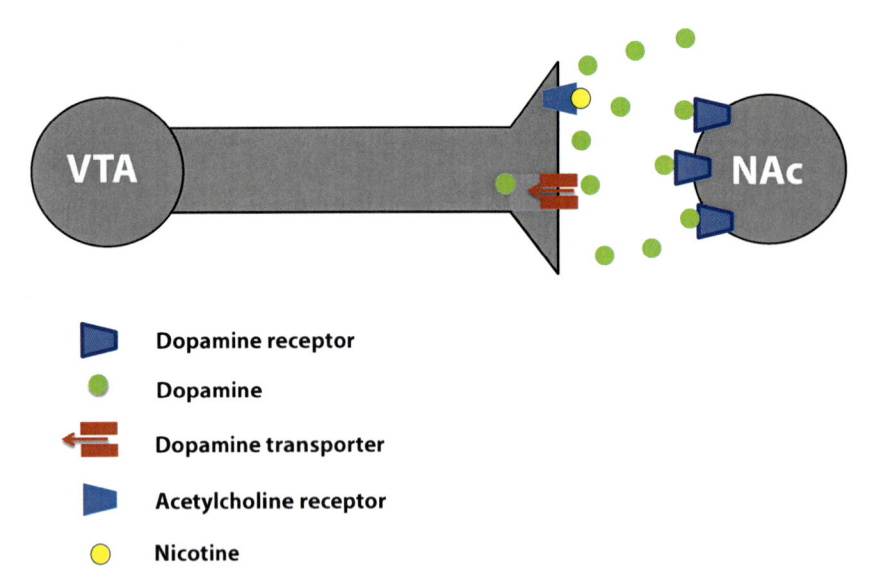

Figure 9-6: Nicotine binds to acetylcholine receptors in the terminals of dopamine-releasing neurons

Opiates, such as heroin and morphine, bind to opioid receptors on inhibitory neurons in the VTA. This causes an inhibition of these neurons. Because the inhibitory neurons in the VTA would normally inhibit dopamine-releasing neurons (Figure 9-7A), the inhibition caused by opiates has the net effect of increasing dopamine release into the synapse at NAc. In plainer terms, dopamine-releasing neurons release more dopamine than usual because there are no neurons stopping them from firing (Figure 9-7B).

GABA: major inhibitory neurotransmitter of the nervous system; made and released throughout the brain

Alcohol induces its rewarding effects through a different mechanism from many other drugs of abuse. It binds to and activates receptors for **GABA**, the main inhibitory neurotransmitter used by the brain (Figure 9-8A). This causes an overall increase in inhibition in many brain areas, causing reduced anxiety and increased motor impairment in the short-term (Figure 9-8B). Like other drugs of abuse, alcohol also increases dopamine release in the brain and is therefore highly addictive.

V. Addiction Treatment

In the laboratory, researchers can use animal models to investigate potential ways to treat drug addiction in people. The model that most resembles human drug addiction is called self-administration,

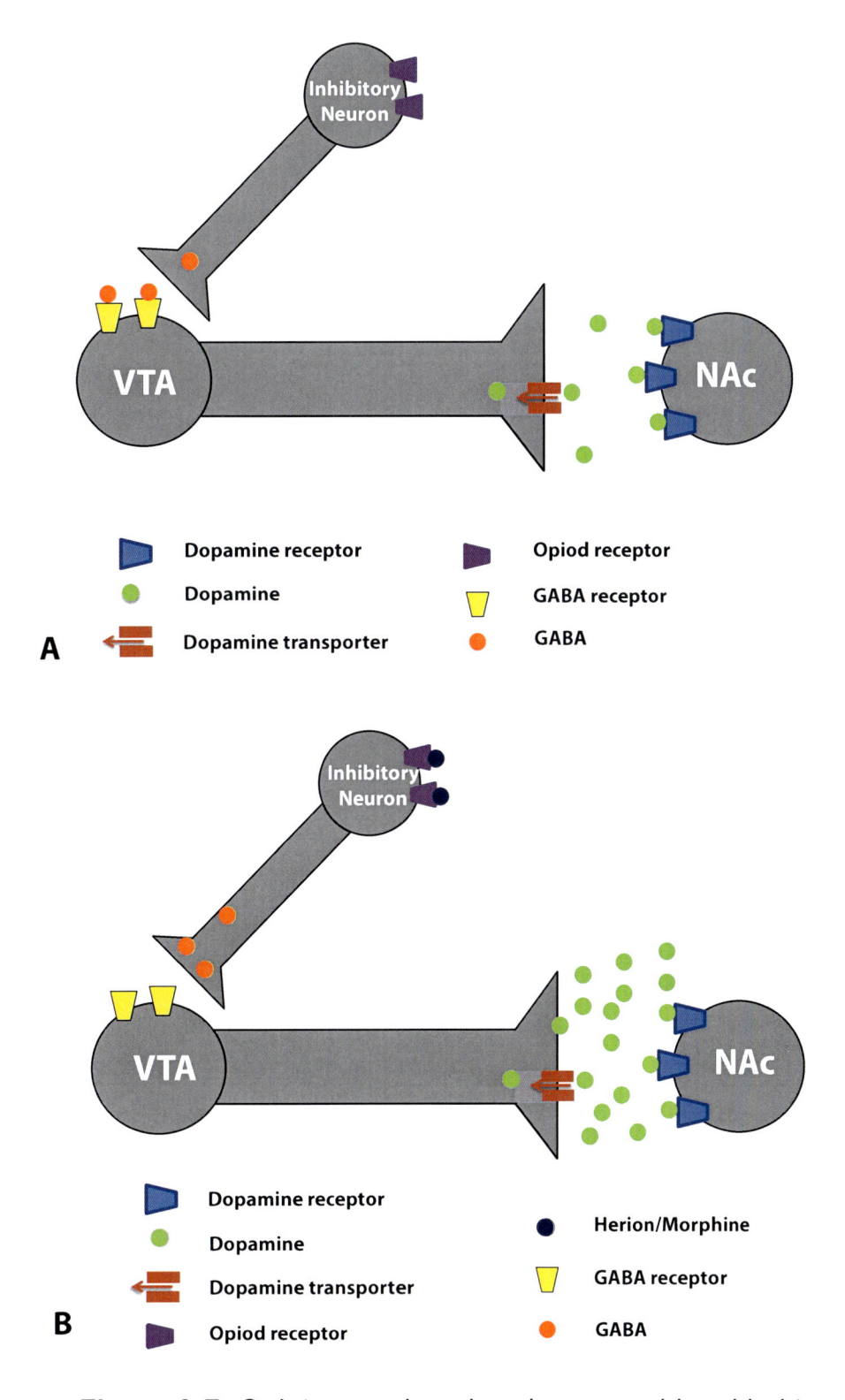

Figure 9-7: Opiates, such as heroin or morphine, bind to opioid receptors, which prevent inhibitory neurons in the VTA from firing

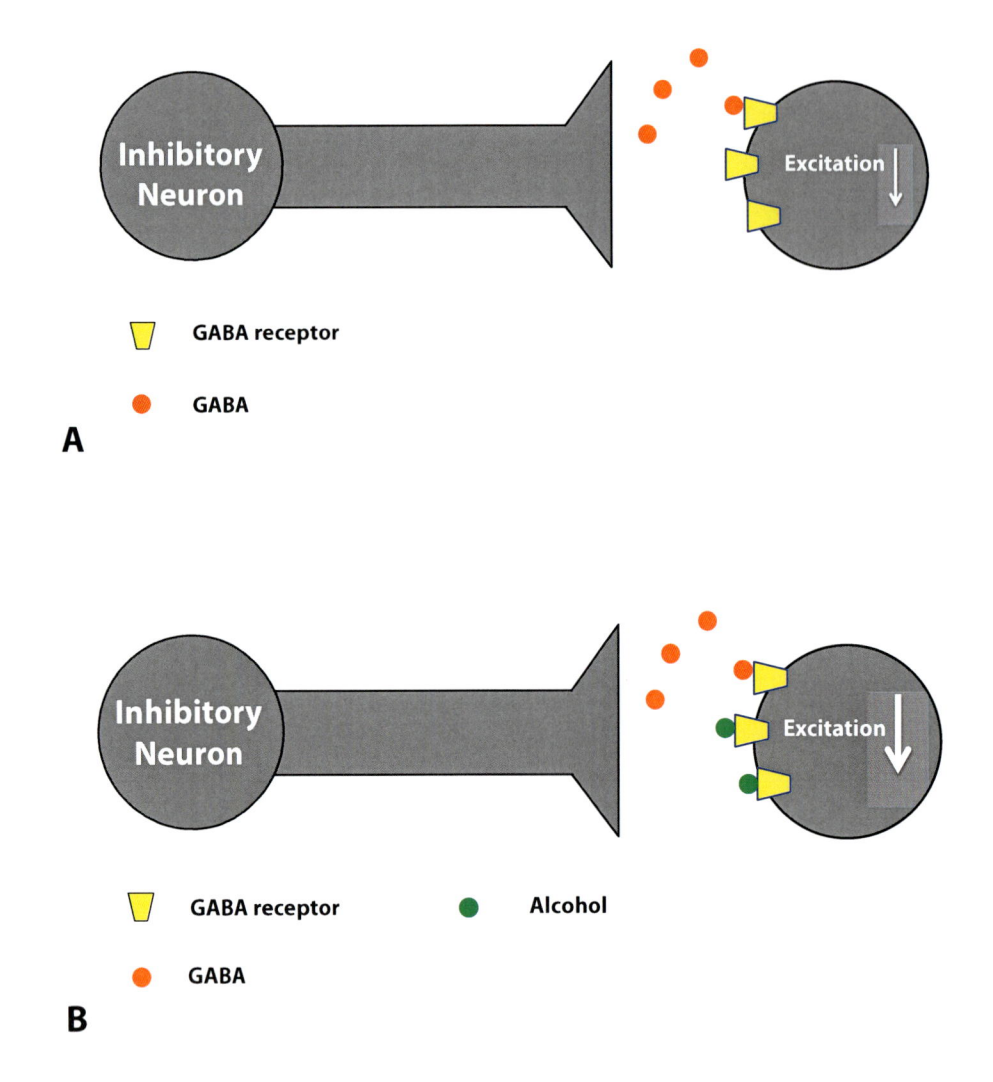

Figure 9-8: Alcohol binds to GABA receptors, increasing the actions of this inhibitory neurotransmitter

in which animals (often mice, rats, or monkeys) are taught to give themselves a drug, usually by pressing a lever. Because animals find the drug rewarding, they will press the lever many times in a row. Thus, researchers can test potential treatments by measuring how much that treatment decreases lever pressing behavior. This research has led to the development of chemicals such as methadone (for heroin addiction) that can help patients stop taking drugs and avoid relapse.

In addition to using chemicals to help reduce craving or prevent relapse, scientists are designing a therapy that may prevent drugs of abuse from ever entering the brain. This therapy relies on the body's own immune system. Normally, the immune system makes antibodies, proteins which attach to pathogens like viruses or bacteria that would make you sick. To treat addiction, scientists have developed an antibody that recognizes and attaches to cocaine instead of viruses or bacteria. This can prevent the cocaine from entering the brain to cause its rewarding effects and promote dependence. This is still in its early stages of development, but scientists hope that the therapy can someday be used to reverse cocaine addiction.

Study Questions

1. What do natural rewards and drugs of abuse have in common?
2. Draw a diagram showing the cycle of addiction.
3. List 3 drugs of abuse and indicate whether they are stimulants, depressants, or both.
4. Explain (in words or with a drawing) how the drugs you named in question 3 lead to increased dopamine release in the NAc.

Challenge Question

Aside from the two examples provided at the end of this chapter, there are many ways scientists might attempt to cure addiction through new drug development. Describe a "drug target" (one of the neurotransmitters or receptors discussed in this or other chapters) that you think could be the focus of a successful addiction treatment. Why do you think this might work, and for which types of addiction would it be most promising?

Brain Byte

Believe it or not, doctors once used the highly addictive opiate heroin to help wean people off of their addictions to morphine (another opiate used as a prescription painkiller). The popular pharmaceutical company Bayer sold heroin for over 10 years as a cough suppressant and addiction cure before it was discovered that heroin is nothing but a faster-acting (and thus more addictive) version of morphine. This is one drastic demonstration of how science and medical opinion must be constantly updated as new information is uncovered!

Made in United States
North Haven, CT
08 June 2023